THE REGULATION OF ASSISTED REPRODUCTIVE TECHNOLOGY

The Regulation Of Assisted Reproductive Technology

Edited by

JENNIFER GUNNING
Cardiff Law School, Wales

and

HELEN SZOKE
Infertility Treatment Authority, Australia

ASHGATE

Published by
Ashgate Publishing Limited
Gower House
Croft Road
Aldershot
Hampshire GU11 3HR
England

Ashgate Publishing Company
Suite 420
101 Cherry Street
Burlington, VT 05401-4405
USA

Ashgate website: http://www.ashgate.com

British Library Cataloguing in Publication Data
The regulation of assisted reproductive technology
1.Human reproductive technology - Law and legislation
I.Gunning, Jennifer II.Szoke, Helen
344'.0953

Library of Congress Cataloging-in-Publication Data
The regulation of assisted reproductive technology / edited by Jennifer Gunning and Helen Szoke.
p. cm.
Includes bibliographical references and index.
ISBN 0-7546-2292-4
1. Human reproductive technology--Law and legislation. 2. Artificial insemination, Human--Law and legislation. I. Gunning, Jennifer. II. Szoke, Helen.

K3611.A77 R444 2002
346.01'7--dc21

2002028165

ISBN 0 7546 2292 4

Printed and bound by Athenaeum Press, Ltd.,
Gateshead, Tyne & Wear.

Contents

PART II: ACCESS TO ART

PART III: ISSUES IN THE REGULATION OF ART

List of Figures

Figure

List of Tables

List of Contributors

Belinda Bennett is a Senior Lecturer in the Faculty of Law, University of Sydney. She teaches health law courses at postgraduate and undergraduate level. Dr Bennett has published extensively on reproductive technology and her book *Law and Medicine* was published in 1997.

Eric Blyth is Professor of Social Work at the University of Huddersfield. He is a member of the British Infertility Counselling Association, the British Fertility Society and the British Association of Social Workers' Project Group on Assisted Reproduction. He is editor of the *Journal of Fertility Counselling* and a member of the editorial board of *Human Fertility*. He has provided consultation on assisted conception for radio and television in the UK and New Zealand and has addressed conferences and seminars around the world. Recent and current research projects include surrogacy arrangements, the welfare of the child in assisted conception, sharing genetic origins information in third party assisted conception, consumers' experiences of assisted conception treatment services, and experiences of women and their partners who have participated in an 'egg sharing' program.

Anne Brewaeys is a psychologist at the Netherlands Institute of Social and Sexological Research. She is involved in infertility counseling, and has also undertaken extensive research involving the follow-up studies of children born by means of new reproductive technologies. She has published extensively in this area.

Leslie Cannold is a multi-disciplinary researcher with degrees in psychology, bioethics and education. She is the author of numerous academic and community-based articles as well as several book chapters and a book on diverse topics including the politics and ethics of abortion, women's moral reasoning and the social and psychological aspect of HIV/AIDS. A regular media commentator on ethical and family issues, she is currently a Fellow at the Centre for Philosophy and Public Issues at the University of Melbourne.

Rosemary De Luca is the Foundation Chair of New Zealand's National Ethics Committee on Assisted Human Reproduction, a ministerial advisory committee. She was until recently Chair of Waikato Ethics Committee, which reviews local and multi-center research projects related to health; was a member of the Ministry of Health's Working Groups revising the National Standard for Health and Disability Ethics Committees; and chaired Waikato's Ethics Advisory Group, which worked with senior management at Waikato Hospital. She has twelve years' experience in health sector ethics and teaches and has published widely in this field. Her current research interests include the language and procedures associated with informed consent to medical treatment in hospitals. Rosemary is a Senior Lecturer in the Department of Arts and Language Education, University of Waikato.

Ruth Deech has been Principal of St Anne's College, Oxford University, since 1991 and was Chair of the UK Human Fertilisation and Embryology Authority 1994-2002. This Authority was established by law in 1990 to regulate the practice of infertility treatment in the UK. It advises government and the public on new issues in reproductive science, licenses all clinics, treatments and research on embryos, and keeps a database on treatments. A lawyer by training, Ruth Deech taught family and property law in Oxford and abroad for many years before becoming Principal. She is an Honorary Bencher of the Inner Temple, a Rhodes Trustee and a Pro Vice-Chancellor of Oxford University.

Giuliana Fuscaldo is a science graduate with a genetics major. She has worked for 10 years in ARTs primarily involved in clinical embryology and research and was responsible for establishing the ICSI program at Monash IVF. As a result of her work in IVF she became interested in the ethical issues arising from advances in reproductive technologies and completed a Master of Bioethics at Monash University's Centre for Human Bioethics. She is now a PhD student at Centre for the Study of Health and Society, University of Melbourne. Her PhD is an investigation into the moral value of genetic relatedness.

Lynn Gillam is a Lecturer in Health Ethics at the Centre for the Study of Health and Society, University of Melbourne, and a Research Fellow in Ethics at the Murdoch Children's Research Institute. She teaches health ethics at both undergraduate and post-graduate level at the University of Melbourne. She is currently involved in a funded project to develop a multimedia package for the teaching of medical ethics. Her academic background is in philosophy and bioethics. Lynn is the founder of the Association for Teachers of Ethics and Law in Australian Medical Schools and a member of the ethics committees of the Department of Human Services and the Alzheimers Association of Victoria. Her research interests include clinical ethics, genetic testing, prenatal diagnosis, research ethics and medical ethics education.

Jennifer Gunning was for three years Secretary of the Voluntary Licensing Authority for Human In Vitro Fertilisation and Embryology. Subsequently she wrote a report for the UK government on human IVF and embryo research and related issues to provide information for Parliament during the passage of the Human Fertilisation and Embryology Bill. She has since written and spoken widely on regulatory issues in assisted reproduction. She is now working as an independent consultant and is a Senior Research Fellow in medical law and ethics at Cardiff Law School.

Jothi Kumar is an Associate Professor and consultant with the National University of Singapore and the National University Hospital. He is Head of the Division of Reproductive Medicine. Associate Professor Kumar provides clinical services at the hospital and is involved with the teaching of medical undergraduate students and postgraduate students.

Francine Manseau is a Senior Policy Analyst with Health Canada and has been working in the areas of policy development and formulation at the federal Department of Health for more than fifteen years. She holds a masters degree in Economics from the University of Ottawa. Her most recent work involves developing proposals for legislation governing assisted human reproduction. As part of that work, she took part in organizing and conducting a series of consultations with various stakeholder groups and individuals. She is also responsible for coordinating the work of the Departmental Advisory Committee on Reproductive and Genetic Technologies. Prior to this, she had participated as key policy advisor in the development of various national public health strategies.

Brenda McGivern lectures at James Cook University (Queensland) in the areas of Torts Law and Constitutional Law. She graduated from Murdoch University with degrees in Law and Commerce and she is admitted to practice as a barrister and solicitor in Western Australia. Brenda's primary area of research interest is Health Law and she is currently completing a PhD in that area through the University of Tasmania. She is a legal consultant with Downings Legal where she works principally in risk management and health law.

Shaun D. Pattinson is Lecturer in European, Comparative and International Law at Sheffield University, UK. He has published widely in the area of medical law and ethics and is the author of *Influencing Traits Before Birth* (Ashgate, 2002).

Guido Pennings is a Senior Research Fellow in the Department of Philosophy and Moral Sciences at the Free University Brussels (Belgium). He obtained his PhD in moral sciences on the ethical aspects of assisted reproduction with donor gametes. Due to his specialization in medical ethics and bioethics he is a member of several ethics committees among which are the National Advisory Committee on Bioethics and the Ethics Task Force of the European Society of Human Reproduction and Embryology (ESHRE). He has published numerous articles in medical and philosophical journals.

Mark W. Richardson is Senior Associate at Blake Dawson Waldron. He practices in the health law area principally on the instructions of Medical Defense Organizations on behalf of medical practitioners. He has a special interest in the increasing interactions between the law and science in the fields of medical ethics and privacy. As well as a law degree he holds an MSc.

Douglas Saunders is Professor and Head of Obstetrics and Gynaecology in the Department of Obstetrics and Gynaecology at the University of Sydney. He is also Academic Head of the Department of Obstetrics and Gynaecology at the Royal North Shore Hospital, NSW and Director of North Shore Fertility, Chatswood. He is the Chair of the Reproductive Technology Accreditation Committee (RTAC) of the Fertility Society of Australia (FSA), the body which accredits infertility clinics in Australia and New Zealand. Professor Saunders has over 300 publications in the area of infertility with a specific interest in the obstetric and pediatric outcomes following ART.

Julian Savulescu is Director of the Ethics Unit at the Murdoch Children's Research Institute, Royal Children's Hospital, Melbourne, Australia. The Unit is devoted to research, education and stimulating open public discussion in the ethical implications of the new genetics. He is also Director of the Bioethics Program at the Centre for the Study of Health and Society at the University of Melbourne. The Bioethics Program provides undergraduate and post-graduate teaching in applied ethics. Editor of the prestigious Journal of Medical Ethics, the highest impact journal in medical and applied ethics, he is also a regular media commentator on ethical issues.

Julian is qualified in medicine, bioethics and philosophy and has published widely in all three disciplines. He was a Clinical Ethicist at the Oxford Radcliffe Hospitals. Julian will take up the Uehiro Chair in Applied Ethics at the University of Oxford in July 2002. This is a newly endowed Chair.

Helen Szoke has been the Chief Executive Officer of the Infertility Treatment Authority in Melbourne, Victoria, Australia, since 1996. The Infertility Treatment Authority is a government statutory authority with responsibility for the implementation of statutory provisions outlined in the *Infertility Treatment Act 1995*. Prior to taking up this position, she was the Executive Manager of the Royal Melbourne Hospital Psychiatry Clinical Business Unit. She has served as a committee member and Chairperson of the Ethics Committee of the Royal Women's Hospital, a committee and executive member of the Victorian Council of Social Services and an inaugural executive member of the Consumers Health Forum of Australia. She is a council member of Women's Health, Victoria.

Helen is currently undertaking a PhD at the University of Melbourne, with the Political Science Department, looking at the policy issues associated with the regulation of reproductive technologies.

Kristen Walker is a Senior Lecturer in Law at the University of Melbourne and teaches in the areas of Constitutional and Administrative Law, International Law, Law and Sexuality, History and Philosophy of Law, Criminal Law and Principles of Public and International Law. She is the Faculty Adviser to the Melbourne University Law Review and has been a member of the Human Research Ethics Sub-Committee between 1995-97. Kristen held the position of Adjunct Professor at Columbia University Law School from September 1998 to May 2000. She was an associate to Sir Anthony Mason, Chief Justice of Australia, from 1993-94. She has many publications in the areas of sexuality, human rights and the law.

Louis Waller (Emeritus Professor, AO, FASSA) held the Sir Leo Cussen Chair of Law in Monash University, to which he was appointed in 1965 until his retirement on 31 December 2000. From 1968 to 1970 he was Dean of the Faculty of Law.

He was Victoria's (Australia) Law Reform Commissioner from 1982 to 1984, and in November 1984 he became the first Chairman of the Law Reform Commission of Victoria. From 1982 to 1984 he chaired the

Victorian Government's Committee on the Social, Ethical and Legal Issues arising from IVF, and in 1985 he was appointed as the first Chairman of the statutory Standing Review and Advisory Committee on Infertility. He became the first Chairperson of the Infertility Treatment Authority in September 1995.

He is also Chairman of the Appeals Committee of the Royal Australian College of Surgeons and was Chairman of the Ethics Committee of the Walter and Eliza Hall Institute of Medical Research from 1995 to 2000.

Professor Waller's main teaching and research responsibilities have been in the areas of criminal law, evidence and law and medicine. He is the author of several well-known textbooks and of numerous law review articles, book chapters and reports.

Acknowledgement

We would like to acknowledge contributions made by Department of Human Services in Victoria, Qantas Airlines and the Infertility Treatment Authority for the significant support and resourcing given to the international symposium.

We would like also to thank the contributors to this volume, many of whom came to Melbourne with major presentations in other areas, some of whom made the trip to Melbourne specially to contribute to the symposium. Their involvement in putting together papers for final publication have made this volume possible and has recorded in history the state of ART regulation in the year 2001.

Helen Szoke
Jennifer Gunning

List of Abbreviations

ACBA	Australian Catholic Bishops' Conference
ACT	Australian Commonwealth Territory
AFA	Australian Families Association
AID	Artificial Insemination by Donor
AIDS	Auto-Immune Deficiency Syndrome
ANZICA	Australia and New Zealand Infertility Counsellors Association
AOFOG	Asia-Oceania Federation of Obstetrics and Gynaecology
ART	Assisted Reproductive Technology
ASRM	American Society for Reproductive Medicine
BSE	Bovine Spongiform Encephalopathy
CDC	Centers for Disease Control (USA)
CECOS	Centres d'Étude et de Conservation des Oeufs et du Sperme
CJD	Creutzfeldt-Jakob Disease
COAG	Council of Australian Government
DHHS	Department of Health and Human Services (USA)
DI	Donor Insemination
DNA	Deoxy-Ribonucleic Acid
EAB	Ethics Advisory Board (USA)
ES	Embryonic Stem (cells)
EU	European Union
FCSRCA	Fertility Clinic Success Rate and Certification Act (USA)
FNA	Fertility Nurses Association (Australia)
FSA	Fertility Society of Australia
GIFT	Gamete Intra-Fallopian Transfer
GMO	Genetically Modified Organism
HCR	Helsinki Committee Regulations
HES	Human Embryonic Stem cells
HEW	Department of Health Education and Welfare (USA)
HFE Act	Human Fertilisation and Embryology Act (UK)
HFEA	Human Fertilisation and Embryology Authority
HIV	Human Immunodeficiency Virus
HREOC	Human Rights and Equal Opportunity Commission
ICSI	Intra-Cytoplasmic Sperm Injection

IEC	Institutional Ethics Committee
IFFS	International Federation of Fertility Societies
IVF	In Vitro Fertilization
MCP	Model Certification Program (USA)
NBAC	National Bioethics Advisory Commission (USA)
NECAHR	National Ethics Committee on Assisted Human Reproduction (New Zealand)
NHMRC	National Health and Medical Research Council (Australia)
NHS	National Health Service (UK)
NICHD	National Institute of Child Health and Human Development (USA)
NIH	National Institutes for Health (USA)
NPSU	National Perinatal Statistics Unit (Australia)
NSW	New South Wales
PGD	Preimplantation Genetic Diagnosis
RANZCOG	Royal Australian & New Zealand College of Obstetricians and Gynaecologists
RMU	Reproductive Medicine Unit
RTAC	Reproductive Technology Accreditation Committee (Australia)
SA	South Australia
SACRT	South Australia Council on Reproductive Technology
SART	Society for Assisted Reproductive Technology (USA)
SDA	Sex Discrimination Act (Australia)
SIRT	Society for Infertility and Reproductive Technology (Australia)
TGA	Therapeutic Goods Act (Australia)
TPA	Trade Practices Act (Australia)
UAE	United Arab Emirates
WA	West Australia

INTRODUCTION

Chapter 1

Introduction

Helen Szoke and Jennifer Gunning

Each year around the world there are meetings of infertility practitioners scientists, counselors, nurses. Included in these meetings are ethical considerations and social considerations about the reach of reproductive technology. The opportunity created by the International Federation of Fertility Societies (IFFS) Meeting in Melbourne also provided the opportunity for the first time to conduct a symposium which would bring together those people involved in the regulation of reproductive technology and those people commenting on the regulation of reproductive technology. This satellite symposium was held in Melbourne, Australia in November 2001. The impetus for the symposium came predominantly from the Infertility Treatment Authority, which is the State Regulatory Authority in Victoria, Australia. Victoria was the place which first passed laws to regulate infertility treatment in 1984. It was the first common law jurisdiction in the world to see the passage of such legislation and followed the review of the new IVF technology conducted by Professor Louis Waller, who was then the Law Reform Commissioner in that State. So it seems timely that the first full meeting to address the regulation of reproductive technology should also be conducted in Melbourne.

The meeting built on many works including the European workshops conducted by Dr Jennifer Gunning looking at the research, ethical and legal aspects of assisted conception. This book brings together the presentations made to this symposium and provides a unique insight into the status of regulatory dialogue internationally in 2001. This time is marked by increasing interest and burgeoning discoveries in the area of embryonic stem cell research; by expanded opportunities and challenges posed by genetic screening processes through the use of preimplantation genetic diagnosis (PGD); through growing understanding about the importance of these technologies where third party conception is used and particularly the issue of access to birth origin information where a donor treatment procedure is used; and, finally, the interface of issues relating to human rights and to community interest and public interest concerns. This is a common theme through all of the issues and all of the applications associated with assisted reproduction.

This publication is organized as follows. The first section provides an overview of Assisted Reproductive Technology legislation around the world. This overview covers Europe with the United Kingdom, Asian countries, Northern America and of course Australia and New Zealand. Missing in the overview is South American continent and the African continent. This gap may be addressed by future organizers of such symposia.

The next collection of issues relates to matters associated with access to reproductive technology. In some regulatory contexts, this is an acute issue where the consideration of individual rights to have treatment may be affected by social regulation which limits access to treatment by certain groups of people including single and lesbian women. These chapters address some of those issues in specific case studies, most particularly in the United Kingdom and in Victoria, but also review the complexity of issues around access and genetic screening services. The area of genetic screening will obviously open up more possibilities and more challenges as it develops.

The final section of the book deals with general issues in the regulation of ART. This is a broad range of issues with a significant focus on genetics and genetic screening but also addressing some of the complex issues related to the use of embryos in destructive embryo research.

This collection is by no means attempting to provide a comprehensive account or overview of all of the issues associated with the regulation of reproductive technology. It is a collection which draws together the views of those associated with assisted reproduction or interested in its regulation which contribute to the body of scholarly debate and review in this contentious area. Just as the last century may have been seen as the nuclear century this century will be seen as the century of bioethics. Reproductive technology is one small component, and is critical in that life is created in unusual circumstances. The very existence of reproductive technologies challenges our traditional understandings of what may have constituted a family, what may have constituted procreation, what is the appropriate age for procreation. It is important to understand all of the implications of the use of these technologies and to share that understanding where possible. From two regulated jurisdictions, one in Victoria and one in the United Kingdom, the editors of this publication have tried to draw together a rich smorgasbord of those issues to inform and encourage further public debate and further meetings in the future.

PART I
ART LEGISLATION AROUND THE WORLD

Chapter 2

Current Legislation In Europe

Shaun D. Pattinson

The enactment of legislation addressing the developing genetic and reproductive technologies has been a slow process in Europe and, indeed, the rest of the world. Even now, almost a quarter of a century after the birth of first IVF baby, many countries have no direct primary legislation. Since this is largely due to the number of differing ethical, political, and religious views, those countries that have managed to pass legislation have introduced compromise solutions reflecting their specific social and cultural environment. Nonetheless, it will be argued that evidence of political compromise and inconsistent values is displayed more clearly on the face of the legislation of some EU countries than that of others.

Although there is apparently widespread agreement across Europe on issues such as IVF, reproductive cloning and germ-line gene therapy,[1] this chapter will focus on three areas on which there is greater disagreement - eligibility criteria for access to assisted reproduction, gamete and embryo donation, and *in vitro* embryo research.[2] It will be argued that many of these regulatory differences can be explained in terms of differing stances on the status of the human embryo and the traditional family unit. Moreover, it will be argued that, in the long term, restrictive regulatory approaches are likely to be undermined by pressures in favor of embryo research and increased social acceptance of non-traditional families.

Overview of current legislation

Despite being widespread throughout the Western world, IVF and its related technologies still evoke considerable controversy. As shown in Table 2.1 below, six of the fifteen EU countries still have no primary legislation explicitly addressing the developing reproductive technologies. Further, what legislation does exist is primarily little more than a list of prohibited activities - a notable exception being the UK legislation. Much of this legislation is directed towards specific issues. The Finnish legislation, for example, is only concerned with medical research, including research on gametes and embryos.

Table 2.1 Current legislation

Country	*Key legislation*
Austria	*Law No. 275* of 1992.[3]
Belgium	None.
Denmark	*Law No. 460* of 1997.[4]
Finland	*Medical Research Act No. 488* of 1999.[5]
France	*Laws 94-653 and 94-654* of 29 July 1994.[6]
Germany	*Embryo Protection Act 1990.*[7]
Greece	None.
Ireland	None. (The Eighth Amendment to the constitution implicitly prohibits some activities: see below.)
Italy	None.
Luxembourg	None.
Netherlands	New legislation (title not known).[8]
Portugal	None. (Some constitutional provisions might be relevant.)[9]
Spain	*Law 35* of November 1988.[10]
Sweden	*Law No. 1140* of 20 December 1984; *Law No. 711* of 14 June 1988; *Law No. 115* of 14 March 1991.[11]
UK	*Human Fertilisation & Embryology Act 1990*; *Human Reproductive Cloning Act 2001*.

The paucity of legislation indicates that many countries are relying on alternative regulatory mechanisms. Legislation is, after all, only one of many possible regulatory responses, ranging from constitutional provision to professional self-regulation. Ireland provides an illustrative example. The Eighth Amendment to the Irish Constitution (which forms Article 40.3.3) states that,

> The State acknowledges the right to life of the unborn and, with due regard to the equal right to life of the mother, guarantees in its laws to respect, and, as far as practicable, by its laws to defend and vindicate that right.

In addition to its effects on the legality of abortion,[12] this provision is widely regarded as implicitly prohibiting genetic diagnosis of the pre-implantation embryo, in vitro embryo research, and some forms of cloning. Additionally, the Medical Council's ethical guidelines declare that certain activities (such as creating embryos for research) constitute professional misconduct, thereby indicating that the Council will revoke the license to practice of any medical practitioner who performs the prohibited activities.[13] Like many European countries, Ireland is considering enacting more specific legislation and the Government set up a Commission on Assisted Human Reproduction in 2001 to explore this area.[14]

Setting up an advisory body is now a standard precursor to the enactment of legislation. Just as the UK's Human Fertilisation and Embryology Act was preceded by Warnock Committee and Germany's Embryo Protection Act was preceded by the Benda Commission, most states considering legislation have set up such advisory bodies. In Greece, for example, a National Bioethics Committee has been established to investigate the issues raised by the application of biological sciences and to investigate their moral, legal, and social dimensions and consequences.[15]

Whether any legislative recommendations see light of day is another matter. Belgium and Italy, in particular, have seen numerous proposed bills fail. The Portuguese parliament even got as far as accepting legislation on medically assisted procreation. Unfortunately, the President of the Republic then exercised his veto, so the legislation never came into force.[16]

Given the level of dissention surrounding the developing genetic and reproductive technologies, it is not surprising to discover that there are a number of issues on which individual countries have taken divergent regulatory approaches. This paper will briefly explore three issues on which there are many different regulatory responses: eligibility criteria for access to assisted reproduction, gamete and embryo donation, and in vitro embryo research.

Eligibility criteria for access to assisted reproduction

The potential of IVF to re-define the family has produced different approaches to the issue of whether access will be granted to those who do not intend to create a traditional family unit. Of those countries with legislation addressing access to assisted reproduction, the Austrian, Danish, French, and Swedish legislation restrict access to cohabiting couples, and only that of Germany, Spain, and the UK appear to allow single woman to gain access. On the face of it, the most restrictive is the French legislation, which effectively prohibits the use of assisted reproduction for homosexual couples, single women, non-cohabiting heterosexual couples, and post-menopausal women.[17] Also, if unmarried, the couple must have been cohabiting for at least two years.

Legislation is not, however, the end of the matter. Many countries without legislation still restrict access to assisted reproduction by non-legislative criteria.[18] In practice, such services are only available to married couples and heterosexual cohabitees in countries such as Finland and Portugal.[19] Additional criteria are also used in countries that have legislation. In Germany, for example, the legislation is supplemented by the guidelines of the Federal Physicians' Chamber, which require physicians to ensure that the couple are in a stable relationship and, usually, married (see Beyleveld and Pattinson 2000, 218).

Even the generally permissive UK legislation requires that account be taken of the welfare of the child who will be born or affected as a result, *including the need for a father* of any child born as a result of treatment.[20] Consequently, even in the UK, it is more difficult for non-standard family units to gain access to assisted reproduction, though no one is legally excluded. Indeed, the statutory duty to take account of the child's need for a father is a clear hurdle for women who are unmarried and not receiving treatment with a man, as any resulting child is rendered *legally fatherless*.[21] Moreover, in practice, many clinics operate additional eligibility requirements, especially for treatment paid for by the state.[22]

Gamete and embryo donation

Differing regulatory approaches are also taken towards gamete and embryo donation. Of those countries with legislation, all allow sperm donation, subject to conditions. Indeed, artificial insemination by donor is now widespread throughout Europe and the Western world.[23]

Legislative positions with regard to oocyte and embryo donation are far more diverse. Only a few countries specifically allow oocyte donation - including Denmark, France,[24] Spain, and the UK - and, of these, all except Denmark[25] allow embryo donation.[26]

Austria, Germany, and Sweden take a particularly restrictive position, prohibiting all oocyte and embryo donation. Under the German legislation, for example, it is a criminal offence to transfer into a woman an unfertilized egg cell produced by another woman, and to attempt to fertilize artificially an egg cell for any purpose other than bringing about a pregnancy of the woman from whom the egg cell originated.[27] In contrast, Spain and the UK once again take a particularly permissive position. Both allow the donation of gametes and in vitro embryos is permitted subject to conditions, such as consent and anonymity of the donor.[28]

The full spectrum of positions can be seen in countries without legislation. In some countries (such as Belgium) both oocyte and embryo donation are practiced, in others (such as Portugal)[29] only oocyte donation is practiced, and in others (such as Ireland) only sperm donation occurs (see Table 2.3, below).

Like restrictive eligibility criteria, one suspects that restrictive regulation of donations of genetic material stems from the value being placed on the traditional family unit. Historically, even donor insemination was thought to be a form of adultery. Since oocyte and embryo donation force distinctions between social, biological, and gestation parentage, it is not surprising to find less than universal support for them.

In vitro embryo research

Differences are also displayed by the regulation of in vitro embryo research, which has proven to be one of the most controversial aspects of the developing genetic and reproductive technologies.

Not all manipulation of embryos is properly considered research. For ease of expression, I will use the phrase "experimental treatment" to cover experimental manipulation of an embryo that is for the direct benefit of that specific embryo. The term "research" will, therefore, be restricted to all *other* experimental manipulations on the embryo. It follows from this that, in Table 2.2 below, Austria and Germany are to be characterized as prohibiting in vitro embryo research, even though they permit experimental treatment on embryos.

Table 2.2 In vitro embryo research[30]

Permitted	*by legislation (subject to conditions)*	Denmark Finland France Netherlands	Spain Sweden UK
	in the absence of legislation	Belgium Greece	Italy Portugal (but not performed)
Prohibited	*by legislation*	Austria	Germany
	in the absence of legislation	Ireland (constitutional provision)	
Unknown	Luxembourg (no legislation and no embryo research is performed)		

As Table 2.2 demonstrates, there is no European consensus on the regulation of in vitro embryo research. Although only Austria and Germany have prohibited such research by legislation, stringent conditions are attached in the seven countries with legislation permitting it. No country, for example, permits embryo research beyond 14 days after fertilization (excluding periods of cryopreservation). Moreover, only the legislation of Denmark, Sweden, and the UK appear to allow the creation of embryos for research (see below). The French legislation is the most restrictive of that allowing embryo research as it only permits research "exceptionally" where it does not impair the embryo.[31]

Again the Austrian and German legislation stand out as being particularly restrictive. On the face of it, like the Irish constitution, they appear to offer maximum protection to the embryo. Nonetheless, the Austrian and German legislation permit the destruction of surplus embryos (i.e., embryos left over from IVF treatment). Whether or not this position is compatible with a strict "pro-life" stance,[32] the Austrian and German prohibition of embryo research can also be explained by a societal need to disassociate from historical acts done in the name of eugenics.

In the long-term, a strictly prohibitive stance towards embryo research is likely to give way to pressures created by economics, scientific prestige, and accusations of hypocrisy or inconsistency. After all, scientists, investment, and seekers of treatment can now easily travel to jurisdictions with less restrictive regulation. Also, in this context, the charge of hypocrisy has potent force, as it is at least *prima facie* inconsistent to reject embryo research as evil yet continue to rely on embryo research being performed elsewhere. Scientific predictions of therapeutic benefits of embryo research can only add to these pressures. Witness, for example, the tensions caused by the realization that embryonic stem cells hold far-reaching therapeutic potential. As a result, restrictive regulation of embryo research is becoming an increasingly difficult position to uphold and the latest regulatory moves all seem to lean towards permissiveness. Only last October, the Netherlands passed legislation allowing scientific research on surplus embryo and providing for a future Royal Decree to allow the creation of embryos for research (subject to conditions). It is envisaged that this decree will be passed in three to five years.[33] Earlier in 2001, the UK extended the purposes for which embryo research is permitted.[34] This year, the French will revise their legislation, which looks likely to produce a more permissive approach.[35] Even the German parliament has recently voted to allow human embryo stem cells to be imported for medical research (see BBC 2002), which seems to be inconsistent with the level of protection given to embryos created in Germany.

Comparing regulatory approaches

As stated at the outset, there are a number of policies explaining the regulatory positions adopted. It is clear that differing values are being attached to the status of gametes and embryos, the traditional family unit, and the potential benefits of assisted reproduction and embryo research. The nature of the balance made varies from country to country.

For ease of comparison, Table 2.3 displays the level of permissiveness of the 15 countries under study. The values given range from the most permissive, denoted by the number 1, to the most restrictive, denoted by the number 3. With regard to *access to assisted reproduction*, 1 means that single women can gain access, 2 that access is restricted to cohabiting couples, and 3 that access is usually restricted to married couples. With regard to *in vitro embryo research*, 1 denotes that it is permitted, as is the creation of embryos for research; 2 that it is permitted but embryos are not allowed to be created for research; and 3 denotes that embryo research is prohibited. With regard to *gamete and embryo donation*, 1 indicates that both are permitted, 2 that oocyte donation is permitted but not embryo donation, and 3 that only sperm donation is permitted. Where there is no relevant legislation or constitutional provision, the *de facto* position is indicated by the number in brackets, which is also displayed where there is a known discrepancy between the strict *de jure* and *de facto* position. Blank spaces indicate absence of information.

Table 2.3 Comparison[36]

Country	*Access to assisted reproduction*	*Embryo research*	*Gamete and embryo donation*
Austria	2	3	3
Belgium	(1)	(1)	(1)
Denmark	2	1	2
Finland	(2)	2	(1)
France	2	2.5	1
Germany	1 (usually 3)	3	3
Greece	(2)	(2)	(1)
Ireland	(3)	3	(3)
Italy	(1)	(1)	(2)
Luxembourg			
Netherlands	(1)	2	(2)
Portugal	(2)	(3)	(2)
Spain	1	2	1
Sweden	2	1	3
UK	1	1	1

Limitations on oocyte and embryo donation are tied up with the value placed on the traditional family unit and the donated genetic material itself. The value attached to the traditional family unit appears to be a key factor, otherwise one would expect all those countries prohibiting embryo donation to prohibit embryo research, which is not the case. As one would expect from countries placing great store in the traditional family unit of a man and a woman raising genetic related offspring, just about all of the countries prohibiting oocyte and embryo donation also have restrictive eligibility criteria for access to assisted reproduction.

A number of countries are clearly camped at polar extremes. Austria, Germany, and Ireland seem to be restrictive across the board. In contrast, Belgium, Italy, Spain, and the UK seem to be permissive across the board.

Some of the countries under study are not uniformly permissive or restrictive. Denmark and Sweden, for example, appear to value the traditional family unit far more than the unimplanted embryo. Both have legislation preventing single woman gaining access to IVF and prohibiting embryo donation, but allowing embryo research. The Swedish legislation goes slightly further in prohibiting the use of donor gametes for IVF. The legislation of both countries also allows the creation of embryos for research. Denmark has, however, now signed and ratified the Convention on Human Rights and Biomedicine without making a reservation to Article 18(2), which prohibits the creation of embryos for research.

At first sight, the French legislation appears to display an apparently inconsistent mix of restrictiveness (stringent eligibility criteria and the prohibition of all embryo research that impairs the embryo) and permissiveness (allowing oocyte and embryo donation, and allowing no-impairing embryo research). Even though the legislation makes it clear that its permissive tendencies are to be regarded as the exception, this legislation is clearly the result of political compromise.

Surprisingly, many Catholic countries, such as Italy and Spain, do not fall within the most restrictive camp. This indicates that there is more to regulatory policy than religion. As indicated earlier, the restrictiveness of Austria and Germany is probably more reflective of their particular historical background than religious conviction.

Conclusion

Not all genetic and reproductive technologies produce widespread regulatory divergence; IVF is permitted in all of the EU countries and all countries with legislation take a restrictive stance towards reproductive cloning and germ-line gene therapy. This paper has, however, focused on three issues on which there is widespread disagreement. Dominant views are still evident - in favor of restricting access to reproductive services to cohabiting couples, in favor of in vitro embryo research, and in favor of gamete donation - but each country seems to have adopted its own socio-political compromise.

There are clear difficulties in predicting the direction of regulatory policy across a large number of countries, even where those countries share many moral and cultural values. I have, however, suggested that the pressures created by genetic and reproductive tourism - the ability and willingness of scientists, investment, and those seeking treatment to travel to other countries - point in the direction of increasingly permissive regulation. In my view, these pressures are particularly powerful with regard to embryo research (including research on stem cells derived from embryos), as even the restrictive countries tend to utilize (and intend to continue to utilize) the insights and developments derived from this research. Moreover, developments in other areas of law across Europe are increasingly recognizing non-traditional family units and the status of non-heterosexual couples.[37] Although some countries are more resistant to change than others,[38] convergence towards permissiveness seems very likely indeed.

Notes

1 Due to space constraints, this chapter does not detail the laws governing cloning and germ-line gene therapy. For these details see Pattinson 2002, especially Appendices 4 and 5.
2 There are also other issues, not covered here, on which there are wide differences among the EU countries, such as the permissibility of preimplantation genetic diagnosis (see Pattinson 2002, Appendix 2).
3 For an English translation see International Digest of Health Legislation 1993.
4 For an English translation see International Digest of Health Legislation 1997.
5 For an English translation see Bulletin of Medical Ethics 2000.
6 For an English translation see Bulletin of Medical Ethics 1994
7 For an English translation see Bulletin of Medical Ethics 1990.
8 The new legislation was passed in October 2001 (see House of Lords Select Committee 2002, para. 7.9). For details see Bulletin of Medical Ethics 2002, 6-7.
9 The Fourth Revision of 1997 contains a number of potentially relevant provisions - Article 26.3 (which guarantees the personal dignity and genetic identity of a human being, particularly in the creation, development and use of technology and in scientific experimentation) and Article 67.2(e) (which places a duty on the state to protect the family by regulating assisted reproduction so as to safeguard human dignity).
10 An English version of this Act appears in *Official Bulletin of the State*, No. 282, Thursday 24 1988. See also International Digest of Health Legislation 1989a.
11 For English translations see International Digest of Health Legislation 1986; International Digest of Health Legislation 1989b; Ministry of Health and Social Affairs 1991, Appendix 2, respectively. Also, Law No. 114 of 1991 regulates the use of certain genetic technology in medical screening (see *ibid.*, Appendix 1).
12 See, e.g., *Attorney General* v. *X* [1992] 1 I.R. 1. A proposed amendment to the constitution, which would have removed the threat of suicide as a sufficient ground for legal abortion, was recently rejected by a narrow majority in a referendum (see Cowan 2002).
13 Section F of these guidelines is devoted to 'Reproductive Medicine'. The prohibition of creating embryos for 'experimental purposes' is in para. 26.2 (see Medical Council 1998, 39).
14 See Irish Department of Health website: http://www.doh.ie, 'Assisted Reproduction Commission set up' dated 25 February 2000.
15 Information provided by Panagiota Dalla-Vorgia.
16 Information provided by Joâo Carlos Loureiro. See also European Parliament 2001, 6.
17 Under Article L-152-2 of the Public Health Code (inserted by Law 94-654 of 29 July 1994) both members of the (heterosexual) couple must be alive, of reproductive age, and married or able to prove that they have been living together for at least two years.
18 For a more in depth discussion of eligibility criteria, see Beyleveld and Pattinson 2000, especially 215–223. This work is, however, now a little out-of-date and my co-author and I did not, then, have access to translations of much of the relevant legislation.
19 Information on Portugal provided by André Pereira. For other countries see *ibid.*
20 S. 13(5), *Human Fertilisation and Embryology Act*.

21 S. 28.

22 The courts have so far refused to interfere with clinics' refusal to treat patients who they consider unsuitable. See, e.g., *R.* v. *Ethical Committee of St. Mary's Hospital ex parte Harriott* [1988] 1 F.L.R. 512 (infertile applicant, with a criminal record for offences related to prostitution, denied access) and *R.* v. *Sheffield Health Authority, ex parte Seale* (1994) 25 B.M.L.R. 1 (37 year old woman denied access to public funded IVF on the grounds of her age).

23 In the absence of legislation, donor insemination is even allowed in Ireland (see Medical Council 1998, para. 26.3, 39).

24 Under Article 152-6 of the Public Health Code, donor gametes may be used as a last resort when medically assisted reproduction within the couple has failed. Under Article 152-5, 'exceptionally' a couple for who medically assisted reproduction without a donor would fail, may receive an embryo.

25 The donation of oocytes is permitted only if they are removed as part of an *in vitro* treatment administered to the donor (s. 14(1), Law No. 460 of 10 June 1997). Whereas the donation of fertilised oocytes is prohibited by s. 7 of Chapter 1, Order No. 728 of 17 September 1997 (see International Digest of Health Legislation 1998).

26 I do not know whether the new Dutch legislation addresses the donation of genetic material.

27 S. 1(1)(i)/(ii), Embryo Protection Act 1990.

28 See s. 5, Law 35 of November 1988 (Spain), and Schedule 3, Human Fertilisation and Embryology Act 1990 (UK).

29 In Portugal only private clinics are using donated occytes and, then, only from non-Portugese women (mainly from Russia). Apparently, Portugese women tend to go abroad for treatment with donated occytes. (Information provided by André Pereira.)

30 This table, deriving from Pattinson 2002, 4.5.2, has been undated for this chapter.

31 Article L. 152-8 of the Public Health Code.

32 I use the term 'pro-life' to refer to the granting of full moral status to the embryo. Supporters of the pro-life position could appeal to a number of principles to defend the destruction of 'surplus' embryos. They could, e.g., appeal to the 'doctrine of double-effect', according to which doing something with foreseeable morally bad consequences can be permissible where it is done with the sole intention of doing something needed to achieve an intrinsic moral good.

33 See the Dutch Ministry of Health, Welfare and Sport website: http://www.minvws.nl.

34 Human Fertilisation and Embryology (Research Purposes) Regulations 2001.

35 The National Consultative Ethics Commission (CCNE) has issued a number of reports recommending allowing controlled use of spare IVF embryos for research, but maintaining the ban on the deliberate creation of embryos for research. A number of sources report that a bill was passed in the National Assembly in January 2002, but is unlikely to be considered by the Senate until after the presidential and parliamentary elections (see, e.g., http://www.laurushealth.com and http://www.katolsk.no). This amendment will allow research on frozen embryos (see Bulletin of Medical Ethics 2002, 6).

36 Information on the situation in practice (in parentheses) was provided by or obtained from Herman Nys (Belgium); André Pereira (Portugal); MacCarthy 2000, 173 (Ireland, donation); and Schenker 1997, 177 (Finland, Germany, Greece, Italy, Netherlands - access and donation).

37 Only last year, e.g., same-sex marriages received legal recognition in Germany.

38 Witness, e.g., the maintenance of the restrictive Irish regulatory position towards abortion in the face of more than 6,000 Irish women travelling to Britain for abortions every year (see Payne 1999).

References

BBC (British Broadcasting Corporation). (2002), 'Germany authorises stem cell imports'. Wednesday 30 January, http://news.bbc.co.uk.

Beyleveld, Deryck, and Pattinson, Shaun D. (2000), 'Legal Regulation of Assisted Procreation, Genetic Diagnosis, and Gene Therapy' in H. Haker and D. Beyleveld, (eds.), *The Ethics of Genetics in Human Procreation*, Ashgate, Aldershot, pp. 215–276.

Bulletin of Medical Ethics. (1990), 'German Law Protects Embryos'. *Bulletin of Medical Ethics,* No. 64, December, pp. 9–11.

– (1994), 'Bioethics Has Become Law in France'. *Bulletin of Medical Ethics*, September, pp. 11–13.

– (2000), 'Medical Research Act 1999 (Finland)'. *Bulletin of Medical Ethics*, No. 155, February, pp. 7–11.

– (2002), 'Stem Cells and Cloning Worldwide'. *Bulletin of Medical Ethics*, No. 174, January, pp. 6–7.

Cowan, Rosie. (2002), 'Irish Reject Tougher Law on Abortion'. *The Guardian*, Friday 8 March, p. 2.

European Parliament (Directorate General for Research). (2001), *Embryos, Scientific Research and European Legislation.* Briefing Note No. 14/2001. (Available at http://www.europarl.eu.int/stoa/publi/pdf/briefings/14_en.pdf.)

House of Lords Select Committee. (2002), *Stem Cell Research.* (London: HMSO) (http://www.parliament.the-stationary-office.co.uk.)

International Digest of Health Legislation. (1986), 'Sweden. Law No. 1140 of 20 December 1984 on insemination'. *International Digest of Health Legislation*, Vol. 37, p. 272.

– (1989a), 'Spain. Law No. 35/1988 of 22 November 1988 on assisted reproduction procedures (*Boletín Oficial del Estado*, 24 November 1988, No. 282, pp. 33373–33378)'. *International Digest of Health Legislation*, pp. 82–93.

– (1989b), 'Sweden. Law No. 711 of 14 June 1988 on fertilization outside the human body' (*Svesk författningssamling, 1988*, 22 June 1988)' *International Digest of Health Legislation*, Vol. 40, No. 1, p. 93.

– (1993), 'Austria. Federal Law of 1992 (Serial No. 275) regulating medically assisted procreation (the Reproductive Medicine Law), and amending the General Civil Code, the Marriage Law, and the Rules of Jurisdiction. Date of entry into force: 1 July 1992. (Bunesgestzblatt für die Republik *Ősterreich*, 4 June 1992, No. 105, pp. 1299-1304)'. *International Digest of Health Legislation*, Vol. 44, No. 2, pp. 247–248.

– (1997), 'Denmark. Law No. 460 of 10 June 1997 on artificial fertilization in connection with medical treatment, diagnosis, and research, etc. (*Lovtidende, 1997*, Part A, 11 June 1997, No. 84, pp. 2195-2198).' *International Digest of Health Legislation*, Vol. 48, Nos. 3 & 4, pp. 321–323.

– (1998), 'Denmark. Order No. 728 of 17 September 1997 on artificial fertilization. (*Lovtidende, 1997, Part A, 26 September 1997, No. 138, pp. 3926–3928)*'. *International Digest of Health Legislation*, Vol. 49, No. 3, pp. 489–491.

MacCarthy, Justin. (2000), 'Splitting the Irish Embryo'. *International Family Law*, pp. 173–176.

Medical Council (Ireland). (1998), *A Guide to Ethical Conduct and Behaviour.* Fifth Edition. Dublin: Medical Council.

Ministry of Health and Social Affairs (Sweden). (1991), *Swedish Act concerning Use of Gene Technology on Human Beings and Experiments with Fertilised Ova.* Stockholm: The Printing Works of the Cabinet Office and Ministries.

Pattinson, Shaun D. (2002), *Influencing Traits Before Birth.* Aldershot: Ashgate.

Payne, Doug. (1999), 'More British Abortions for Irish Women'. *British Medical Journal,* Vol. 318, p. 77.

Chapter 3

The HFEA - 10 Years On

Ruth Deech

Regulation initially arose from the widespread fears expressed by politicians, the media, the public and the professions about embryo research, its morality and its direction. Subsequently, the spread of IVF treatment and clinics came to drive regulation and perhaps only then because the National Health Service has declined to fund the generality of such treatment, giving it the potential to become big business with all the dependant drawbacks thereof. After ten years, the emphasis has reverted to embryo research, its morality and its direction. It is perceived as more important but more unclear than ever before. The funding of infertility treatment in Britain and the ethical debate are complicated by the fact that many do not regard infertility as an illness, even although the pain, distress and widespread nature of it is plainly recognized.

The path of the history of the regulation of IVF in Britain moved from control of embryo research to regulation of infertility treatment, necessitating the organization of centers offering fertility treatment, a code of practice for their behavior, collection of statistics and the licensing of research along publicly recognized guidelines. In recent years there has been a proliferation of new techniques of treatment for infertility requiring consideration of their desirability and safety: intracytoplasmic sperm injection with sperm or spermatids, sorting of sperm by sex, egg freezing, storage of testicular and ovarian tissue, and advanced forms of embryo culture.

There has now emerged an extension of the availability of assisted reproductive techniques in cases of infertility to matters of convenience and preference, for example, the insemination of older women past the age of menopause, posthumous insemination, choice of sex of the baby, egg freezing with a view to postponement of childbearing for personal reasons and insemination of single women or gay couples. Another new area is the globalization of assisted reproductive techniques: gametes and embryos may be imported and exported from country to country in search of one that permits the desired treatment or allows the chosen gametes to be used.

This has sometimes been termed "procreative/reproductive tourism". For example, if a couple in Sweden seeks donor insemination but do not wish to be subject to the Swedish regime of identified sperm donors, they might choose to come to Britain for treatment with anonymous sperm donation. If a woman wishes to be sure that her child will know the name of the donor father, she may leave Britain to go to the States or ask for permission to import a special sample from the US. If a couple are not allowed treatment with spermatids in Britain, or sex selection, they may go abroad to a country that does permit these techniques and return home for the birth. The morality of the profit made by private doctors giving treatment which results in a complicated birth, with costs to be borne by the National Health Service, has yet to be debated but has become a real issue.

The newest trend in ART is the attempted elimination of disease and the addition of desirable characteristics to babies by use of preimplantation genetic diagnosis (PGD), aneuploidy screening and the prohibited techniques of cloning and genetic manipulation. Preimplantation genetic diagnosis is used at the moment only by couples with a significant risk of passing an inheritable disease on to their babies, so significant that it worth undertaking the technique of IVF, with its very low pregnancy success rate, to ensure elimination of the diseased genetic trait. PGD involves the creation of an embryo in vitro for a couple who are not otherwise infertile, with the aim of examining the embryo at the 8-cell stage, that is a day or two after fertilization, to see if it carries the defective genes. A defective embryo will be allowed to perish, but one that is tested clear of the inheritable disease will be transferred to the womb. It may even be that a couple select the defective embryo; it is said that deaf couples may wish to ensure that they have a deaf child, representing to them the normality of this condition and the desirability of its acceptance.

Further genetic manipulation is illegal under British law, as is reproductive cloning, but the issues are worth discussing because they are currently coloring the public's view of the general scientific techniques involved. It was the morality of embryo research that sparked off the debate that ultimately led to legislation in Britain in 1990. The morality of embryo research is by no means a settled question, at least as a matter of public debate and ideology, although its legality is fixed. Although Britain is not a religiously polarized country, and has very few Catholics in the population relative to the US, nevertheless recent public discussion about cloning and associated cell nuclear replacement research has revealed that there is still significant opposition across all religious and social divisions to any form of embryo research, on the grounds that the embryo possesses the full moral status of a human being. If anything, the sensitivities are greater than they were 10 years ago, but both those who were in favor of embryo research, and those opposed to it were able to agree that if it were to be permitted it should be regulated. The status of the embryo and the law that applies to it were settled in the Human Fertilisation and Embryology Act 1990, and that issue has not been reopened. It is salutary to recall now that in the 1970s and 1980s in Britain, embryo research came close to being totally prohibited and that scientists may well have accepted legal regulation unwillingly as the price of being able to continue their work at all. It may be that this history has repeated itself in the closely restricted permission given early in 2001 to open up embryo research for the purpose of the prevention and cure of serious disease. It is unlikely that the regulations would have been adopted by such real majorities in the House of Commons and the House of Lords had it not been for the assurance given that the HFEA would continue to regulate and monitor this new research as strictly as it had done the old. The most severe penalties contained in the 1990 Act pertain to embryo research, which is hedged about by restrictions needing alteration by Parliament, by external control over the scientists of a type not found in any other field, and by criminal penalties for infringement of the permitted procedures. This strictness turned out to be of value in the struggle by scientists to gain public and parliamentary permission to extend embryo research in late 2000.

Legal regulation in Britain has probably served to protect clinicians and scientists not only from legal action for malpractice (where they have followed the HFEA Code of Practice) but also to give them a shield against accusations of ethical malpractice, for they are acting within the parameters agreed by Parliament and the HFEA. These safeguards clearly do not apply where there has been a total breakdown in the clinic and the deliberate flouting of the criminal law, which has happened once or twice. Nevertheless, the excitement of pioneering has not vanished from this area. Those of us who are involved in regulation, as well as the professionals in the field, are very well aware that they are privileged to be operating at the outer limits of technology in a field uniquely intimate and religious as well as scientific, that they are probing the depths of human origins in the face of opposition from many quarters, and that their decisions and discoveries will raise more questions, both religious and secular, and open up old anxieties.

While the Human Fertilisation and Embryology Act is a very satisfactory and flexible piece of legislation, it will not remain the last word, and the ebbs and flows of public opinion on the matters it covers can be felt daily. It has, on the whole, been a triumph of legislative drafting, with more flexibility and discretion than would have been realized in 1990.

But is now 10 years old and there is no doubt that it needs amendment both obviously in relation to reproductive cloning, which it is intended should be banned, and other matters. The new Human Rights Act in Britain has cast doubt on the justice of the system of internal appeals within the legislation, that is, a clinician whose license renewal is refused or limited, is granted an internal appeal to the HFEA. This needs a fresh look. It is also the case that the very strict confidentiality legislated for in 1990 has proved an impediment to much needed follow-up studies on children born, for example, by ICSI. It is urgent that these confidentiality restrictions be removed to allow epidemiologists to carry out the work which alone can give us the confidence that IVF techniques are safe.

The changes in regulation and reproductive science of the last 10 years must be seen against the background of enormous social change in attitudes to the family. Family formation has changed swiftly in the last decade. Feminists have posed their own questions and anxieties in relation to the development of ART. The discoveries made have been so profound that in a sense the parable of Adam and Eve is being re-enacted. The medical advances have been especially rapid since the mid-1990s, and there have been significant legal changes in Britain that have affected the HFE Act. The Human Rights Act was introduced in Britain in 1998 and implemented in 2000. Its promises of the right to respect for private and family life and the right to marry and found a family will no doubt have significant impact on reproductive medical law. Combined with the relaxing effects of European Treaty law, which is also now felt in this country far more than it was 10 years ago, there may yet be an extensive undermining of the ethical and legal controls embedded in English law, when it has to yield to European law or contrary human rights interpretations. Finally the political background has changed. Governments and the public have now come to realize the dangers and benefits, and their immense scale, attaching to genetically modified food, genetic engineering and increased knowledge about the origins of mammals. Reproductive science has become fused in the public's mind with BSE, CJD, GMO, other similar acronyms and the Human Genome Project. In other words, although British education in these matters is not good, the public is aware and very often deeply anxious about changes and the damage that they can do. The government responded to these anxieties by conducting a review of all the committees that offered scientific advice to the government, and rearranging them. Some were abolished but most were regrouped under three umbrella committees: (1) the Food Standards Agency, (2) the Agriculture and Environment Biotechnology Commission and (3) the Human Genetics Commission. In the case of the HFEA, its independent existence has twice been vouched for in Quinquennial Reviews. Because it has its own Statute, it has duties and an independent existence and cannot be subject to the deliberations of the Human Genetics Commission, a non-statutory body. Nevertheless it will listen to its advice and opinion.

The social, medical, legal and political changes of the last 10 years and their impact on the future will now be examined in turn.

Social matters

The deployment of assisted reproductive technology (ART) affects assumptions which we bring to the understanding of family life and also to the very understanding of family life itself. It goes to the heart of our beliefs about the family, marriage and humanity. Every announcement of a new development in the techniques of assisted reproduction is accompanied by an element of publicly expressed fear, as well as elation at the happiness it might bring. That fear relates to the maintenance of the family as we know it with its own power structure and ability to influence the future. On the one hand, the enablement of birth to an otherwise childless couple represents the creation of a traditional nuclear family. On the other, the ability of gay and single women of all ages to conceive, and the use of ART for convenience purposes, has led some to worry about the creation of family forms that are widely disapproved, and a feeling that acceptance of them is being imposed. With the introduction of genetic engineering, there is a popular fear of loss of reproductive choice, as well as its extension in ways that are unpopular. Freedom to reproduce as one wills has not, however, been a universal principle of civilized society. There have always been restraints, designed to achieve certain fundamental purposes. The age of consent for sexual intercourse and to marriage, the number of children regarded as the optimum, arrangement of marriages, prohibited relationships, tolerance of cohabitation, contraception, abortion, adultery and homosexuality, attitudes to female employment outside the home and to childcare, have all fluctuated from time to time and from society to society, as our major influences on the shaping of the next generation. In a democratic society most or all of these factors have been regulated by law. So ART too takes its place as a major factor in reproductive abilities, as important in shaping the family and as necessary to regulate as other factors.

For some decades now there has been increasing tolerance of a high divorce rate, cohabitation, homosexuality, single parenting, abortion, adultery, under-age sex and pregnancy, births out of wedlock and diverse lifestyles. Yet the law about regulation of ART is largely premised in its formulation on help to the traditional nuclear family, which lacked only children to make it complete. Its original success lay in completing those families and, to this day, the underlying assumption of the professionals and the media in handling such cases is that all married women want a baby and that their lives can only be fulfilled and occupied in so doing. Once ART was extended to unconventional families there were complaints that family life was being undermined, even though natural fertility is leading to far more unusual families than ever ART could.

There is a circular relationship of ART and attitudes to marriage and the family. It is tritely observed that new family forms such as one parent families, cohabitation, step families and same sex couples have become more common, as have later marriage and age of childbirth, that there is a decline in marriages and a rise in divorces and remarriages. It is also the case that it is these factors that have driven reproductive technologies forward, as much as the other way round. For ART itself has contributed to an atmosphere where legal and biological parenthood are not taken to coincide and it has provided the facility for the non-traditional to copy the traditional.

The new emphasis on the "family" as distinct from legal marriage has turned away from the traditional two-parents-and-children model and focused instead on the presence of children and the intention to have them. Now it is said that children make a family, rather than the other way round. Hence it must seem to non-traditional units such as cohabitants, same sex partners and single people that the presence of children will make that unit into a recognized family, even if the law refuses that status or society generally does not regard them as such. An example is the recent account of two British gay men who fathered twins by an American surrogate mother from mixed sperm. This led to a public debate as to whether this was achieved out of desire for the children, or desire for the hallmark of a "real family" with children.

The presence of children has been regarded by judges and family lawyers as the key to the application of protective law, almost equivalent to that applying to the married. Thus children have replaced the marriage contract, the wedding ring and certificate as the seal of a "family". These attitudes are bolstered by studies claiming that the traditional two-parent family is not the only satisfactory structure for raising children. There is a possible risk to women in this child-centered approach. The excessive concentration on having children at all costs, as if neither the woman nor the couple can be valid without them, is detrimental to her health and her status. It should be counterbalanced by more concentration on the personalities and contributions of individuals themselves and not as members of a unit or as parents. It should also be recalled that after 10 years of regulated practice, the success rate (i.e. a "take home baby") of all methods of IVF is still only around 18 per cent; 82 per cent of patients will go home disappointed and we need to concentrate more on consolation and respect for them, as well as the voluntarily childless.

A view of the complex interaction between ART, marriage and contemporary notions of the family may be summarized as follows: traditional notions about marriage gave rise to great pressure to have children, and infertility was seem as bad from many perspectives. This attitude encouraged investment in and acceptability of ART, which in turn contributed to the growing rejection of marriage as the only valid union. The newer style unions have sought legal and social validation by recourse to ART to give them children.

The extent and reasons for infertility have some bearing on these issues. It is said that one sixth of couples are infertile, to the extent that they seek medical assistance. This may be because of higher rates of abortion, sexually transmitted diseases[1] and environmental factors in society: as they rise so will recourse to ART. This has come as a surprise to the 1960s generation of women who were the first to believe that they could control their fertility (because of legalized abortion and access to modern contraception), and choose fertility/infertility at almost any stage of life at will. As more of them and later generations of women entered higher education and sought careers, they have postponed childbirth to older ages, finding a shortage of childcare facilities, a lack of family-friendly policies in the workplace, and that men are still unwilling to shoulder their fair share of the burden of parenthood. Divorce and remarriage have also led to older childbearing and much infertility is now due not to pathological causes but simply to aging. As is well known, there is also a shortage of babies available for adoption and thus no alternative. There is little information given to young people on the prevalence of infertility, certainly when compared to the amount of information given about control of conception.

In consequence of these social developments, ART has spread from the treatment of infertility to the overcoming of inability to conceive for practical reasons. It can now assist a woman who has no partner; a widow; homosexuals and lesbians; men who wish to reverse vasectomies; women who wish to reverse the effects of sterilization; HIV; and persons with a genetic disease which they wish to avoid handing on. Where the necessary cure is surgery or drugs, they are outside the remit of the HFE Act.

Women's autonomy is protected to some extent by the 1990 Act and the general medical law of the country, shortly to be reinforced by human rights law. Women run the risks of being regarded as incomplete if they have no children, whether voluntarily or involuntarily. They are also vulnerable to the risk that new techniques of ART, which have scarcely been tested as safe or viable, may be rushed into use by clinicians in order to be ahead of the field and advertise themselves as such. In Britain, we wonder whether it is acceptable for a doctor to encourage a woman to undertake 10 or more treatment cycles, and whether he has had regard to her feelings about this and the effect on the child, were it ever to be born, or existing children. Statutory regulation requires that doctors and scientists inform patients and then abide by the choices that they have made. The rush to fertility treatment, almost at any cost, over the last 10 years has thrown up its own dangers, both to health and to women in general. We must beware the dangers and disappointments of the unlimited pursuit of success in infertility treatment. We must ask whether the passionate desire to become a parent justifies taking of risks to health on a scale that would not be acceptable if the issue were food safety or environmental preservation. Sooner or later, our judges will face questions such as – what if cloning were the only way to enable a particular young woman to become a mother?

Medical developments

Most of the advances in reproductive techniques of the last 10 years are well known, but some comments will be made about some of the more interesting against their social background, and with future implications. A significant situation from the woman's point of view in recent developments in assisted reproduction is the new-found ability to freeze eggs. For many decades it has been possible safely and successfully to freeze sperm, both human and animal. Indeed, scientists allege that sperm may be frozen for many decades without harm to it or to the resulting offspring. It is for ethical and social reasons that there are limits of 5-10 years on storing sperm and embryos.[2] This facility means that a man who faces chemotherapy and radiotherapy cancer treatment, which is likely to affect his fertility, may store his sperm before that treatment begins and preserve it for future successful treatment of his partner. The same possibility was not open to women until very recently.

Early in 2000 the HFEA announced the decision to allow clinics the carefully controlled use of frozen eggs in fertility treatment. This means, for example, that a woman who risks becoming infertile following cancer treatment could also use the eggs that had been frozen before her treatment. The HFEA had at first allowed clinics only to freeze and store human eggs, but not to use them in treatment. This was because research had shown that it was possible to freeze eggs but that the great majority of them did not survive the thawing, that there were genuine concerns about the safety of babies that might be born from frozen eggs and that, in any case, the chances of a successful pregnancy following the freezing of eggs was only around one per cent of treatment cycles. Subsequently the HFEA decided to allow frozen eggs to be used in treatment because the most recent expert evidence had shown that the procedure was sufficiently safe, although still not very successful.

The significance of this was that it was decided not to exclude women freezing eggs for social reasons, as distinct from reasons connected with cancer treatment. Given women's pursuit of careers and the impact of divorce, the female reproductive lifespan appears restricted in comparison with women's lifestyle plans. Frozen eggs might be an attractive option to women who are aware of the problems and concerned to preserve maximum fertility for the time when it is convenient and when they have found the right man. The HFEA avoided claims of medical discrimination of human rights by not limiting egg freezing to oncology patients or those who are losing ovarian function. In permitting the use of frozen eggs, it was also decided that it was unfair to make a distinction between men using frozen sperm in any way that they liked, which is legal, and women using frozen eggs.

The HFEA has only recently moved into a difficult field where public health and safety have to be considered in opposition to the ambition of some clinicians to improve their success rates in infertility treatment, that is the reduction of multiple births. They are dependent on the number of embryos placed in a woman in any one treatment cycle. The HFEA Code of Practice used to limit the number to three, even though the professional guidelines of the British Fertility Society and the Royal College of Obstetricians and Gynecologists recommended a maximum of two. The HFEA wished to leave some flexibility to clinicians but eventually, in the light of the success rates equalizing as between the replacement of two and three embryos, and the increasing desire on the part of some clinicians to go as far as the law would enable them, it reduced the maximum to two. Even when a woman is older than the average, there is a significant risk of her undergoing a multiple birth when three embryos are replaced. Multiple births are associated with premature birth, low birth weight, a higher rate of stillbirths and neonatal death and long-term disability such as cerebral palsy. There is also an increased risk to the mother, including complications at the birth, financial and emotional stress and exhaustion due to coping with more than one baby. In one recently reported year, 47 per cent of individual babies born from all types of IVF came from a multiple pregnancy, and this is repeated worldwide. In fact, so much is it so, that some clinicians and informed persons assume almost automatically that twins or triplets must have resulted from IVF.

In a few exceptional cases it will be permitted to replace three. An example could be an older woman who may have had repeated IVF failures or who has produced three lower quality embryos. These exceptions will be left to clinical judgement, and the HFEA will be monitoring the overall multiple birth rate at the centers and looking at the exceptions detailed in the patients' notes. We certainly recognize the pain that infertility can bring and that the desire to have a child is a very strong one, but it cannot be at any cost. There are already some clinicians who claim that the use of one embryo will one day be the ideal.

Modern treatments for cancer have presented interesting infertility conundrums. In 1997 there occurred the case of a two-year-old boy called Oskar, who was faced with emergency treatment for cancer that might destroy his future fertility. The surgeon suggested the removal of a small piece of his testicular tissue with a view to performing an experimental preserving procedure. Oskar's testicular tissue would be frozen until he was an adult, when it would be possible to reimplant it into him and hope that normal sperm production would be restored, or that by then there would be other as yet undiscovered methods of restoring his fertility. The question was whether the entire procedure was legal. Certainly Oskar's parents could consent to the operation that would remove his testicular tissue because parents may consent on behalf of minors to operations. The problem was that under the Human Fertilisation and Embryology Act no one but Oskar himself would be able to consent to the *storage* of his gametes and that would have to be in writing. By the time he was old enough for reimplantation to be considered, he would be able to sign his own consent to further procedures. The problem was not a real one, however, for medical evidence showed that the testicular tissue of a prepubertal boy did not contain gametes, and would therefore not be subject to the provisions of the HFE Act until it did. It could therefore be stored on the consent of his parents, but restoration would need Oskar's consent. The Authority went on to consider the parallel situation with girls. It ruled that if ovarian tissue were to be removed and stored, it would fall under the provisions of the Act only if it contained an egg, and that this was never the case with prepubertal girls. In the case of women it would depend on whether the particular piece of ovarian tissue happened to contain an oocyte or not. Clearly, the preservation of tissue against the possibility of restoration or new developments in fertility in the future is a big issue. The HFEA is working together with other relevant organizations in the field to cover this.

The same problem arises with stem cells. While the use of embryos and their creation for research is firmly under the remit of the HFEA, the resulting stem cells are not under the remit of any body at the moment. It is expected that parliament will make provision for this, with the advice of the forthcoming report of the House of Lords Select Committee on this topic. In other words, the physical products of fertility treatment are presenting storage and use problems that were not considered in 1990.

In 2000 Britain was engaged in a deep debate on whether it was moral or practicable to extend the purposes for which embryo research may be permitted. In the end, after a passionate debate in the House of Commons and the House of Lords, a large majority in each House approved new regulations which came into force on 31 January 2001 specifying additional legal purposes for embryo research

- increasing knowledge about the development of embryos,
- increasing knowledge about serious disease, or
- enabling any such knowledge to be applied in developing treatments for serious disease.

In response to the new regulations the HFEA has enhanced its existing licensing system. Applicants to do this research will be required to justify why embryonic stem cells are to be used, rather than adult stem cells or animal stem cells. They will be required to provide detailed information on the fate of the stem cells throughout the project, keeping them within the license project. Six monthly progress reports will have to be submitted to the HFEA. Additional guidance has been provided to peer reviewers and new ones with expertise in the field have been sought. Three clinics are currently licensed to generate stem cells under the terms of the current research licenses, and two more have applied to join them.

Although the monitoring will be a detailed and sensitive task, it is fundamentally no different from that which has been carried out by the HFEA over the last ten years. We consider it all worthwhile because of the enormous potential of stem cells as a source of new tissue for therapeutic uses in the repair of damaged tissue and organs for a wide range of currently incurable disorders. Work in animals and early work to extract stem cells from human embryos supports this position. At present stem cells from embryos appear to have the greatest potential to be developed into the widest range of tissues. In the long term, but not in the short term, the scientific view is that it will be possible to program adult cells to make them behave like stem cells with the full potential of embryonic stem cells but without the morally more contestable need to create an embryo. Ethical opinion on the use of embryos in research as a source of stem cells is divided, but the view seems to have prevailed in Britain that it is more ethical to use scientific knowledge to relieve existing human disease than it is to block this research, especially when in the alternative embryos not required for infertility treatment may be allowed to perish, or scientists may go abroad to unregulated countries to carry out research forbidden in Britain. A remaining international problem regards lack of control over the stem cells themselves. Once isolated, they are not subject to the HFE Act which does not cover them, and they could be imported from overseas without the need for regulatory control or, indeed, exported.

Quite where this will lead remains to be seen. It does seem though that the trust placed in the HFEA to monitor and regulate the use of embryos is the foundation of Britain's success in the stem cell research potential fields and that the extension of embryo research again owed something to confidence placed in the HFEA. It is likely that provision will have to be made for the guarding of stem cell lines and that there will be growth in the biotechnological industry on the back of this.

It is thought that the demand for preimplantation genetic diagnosis will increase and the public consultation that was held on this topic did not reveal massive antipathy. The scope of disorders that can be identified is likely to widen. PGD offers hope to many, but that is not to say that PGD is without problems. There is the possibility of misdiagnosis, either because of the techniques involved or because the biopsied material is not typical of the embryo. It runs the risk of reducing the number of available embryos since some may be damaged during the biopsy which in turn reduces the live birth rate below that of a normal IVF cycle. The clear message was given by the public that PGD should not be used to deliberately choose desirable characteristics and also that the use of the technology should not further disadvantage disabled people now and in the future. It is necessary in future to tread a fine line between an outright ban on the use of PGD and total freedom to use it for any and all reasons. Therefore the HFEA, in consultation with the Human Genetics Commission strongly supports the principle that the use of PGD should be limited to specific and serious conditions.

The HFEA is now moving towards widespread licensing of PGD. This will involve insisting that clinics make arrangements for genetic counseling and set out their decision-making process used to determine whether PGD should be offered, as well as the information to the patients.

It is likely that three types of PGD will be licensed: (a) specific diagnoses, including tests for chromosomal aberrations; (b) sexing for X-linked disorders and (c) special categories where the disorder affecting a family does not have a well-recognized pattern of inheritance. This controversial issue will be brought under close regulation with procedures modeled on existing ones, but there will be an even stricter need to satisfy the patients, the public, clinicians and, of course, future generations.

Another issue that is on the horizon is aneuploidy screening. This may be offered to enable clinicians to choose the embryos most likely to result in a viable pregnancy. Older couples in particular are concerned about the risk of Down's Syndrome and all are offered prenatal testing as a routine part of pregnancy care. It is argued that if the possibility justifies *prenatal* testing, amniocentesis and termination, it is reasonable to take alternative steps to exclude chromosomal abnormalities *before* implantation. Thus we come ever closer to the fine line between excluding serious disease, deemed permissible, and choosing "designer babies". Aneuploidy screening may be as far as one can go, justified as it is by increasing the chances of fertility and avoiding miscarriage.

The newest problem to be faced is that of patients' desire to choose an embryo that will give them a desired second child, but one that also has the capacity to donate bone marrow or stem cells to enable an older brother or sister to recover from a genetic disease. This is particularly problematic as it may seem that the second child is being brought into the world solely to help the first, and the public is only likely to feel comfortable with this procedure if satisfied that the patients desired the second baby in its own right, and not just as a means to an end to assist the first. It can also be argued in favor of selection that an embryo is being selected to bring a second child into the world who will not suffer from the same genetic disease as the first.

It may also be commented that follow-up on the children born of these new techniques is more crucial than ever and that the possibilities of this may not have been considered in 1990. As commented above, the confidentiality aspects of the Act stand in the way of this. There is an essential need for the Department of Health to organize and fund studies because only these studies will, in the end, establish the public's confidence in the safety of the new procedures. The balance between the preservation of confidentiality and the assurance of help is as delicate here as it is in the other new field of the use of genetic databases.

To summarize: the newest techniques to have emerged all require follow-up studies; some of them will require a further organization or an extension of the HFEA to safeguard resulting stem cells or tissue which need to be protected against misuse and kept safely; there are issues of whether one is going too far towards the feared concept of "designer babies" in some of these new techniques; and the use of those that prolong fertility or, in the interests of public health and safety, constrain clinicians, raise issues of lifestyle and human rights which may become important issues before British judges. In this area frozen eggs, sex selection and number of embryos to be used might all spark off a human rights challenge.

Human rights

Reference has already been made to the enactment of the European Convention on Human Rights in Britain. It requires that all legislation be interpreted as far as possible in a way which is compatible with the Convention.[3] Relevantly to the HFEA, Section 6 of the Human Rights Act 1998 requires public authorities to act in compliance with the Convention unless they are prevented from doing so by statute; thus an ambiguity in a statute has to be resolved with reference to human rights principles. Failure to comply with human rights obligations is a ground of illegality in judicial review proceedings brought by alleged victims, thus forcing public bodies to apply human rights in all aspects of their work, save where there is a statutory obligation pointing the other way.

Article 8 protects private and family life, home and correspondence. This Article might be relied on a woman such as Mrs. Blood seeking insemination with her late husband's sperm, regardless of how it was procured; it might also be relied upon by an infertility patient who felt that her treatment was being unduly delayed because of the shortage of gametes in this country and a disproportionate refusal of permission to import sperm from abroad. In the alternative, she might want cloning. Therefore the questions to be asked are whether access to insemination by a widow amounts to private life or family life and whether the desire by an infertile person for a baby by any means is included. There is no doubt that the effect of human rights law will be to cause the NHS in Britain to extend and level out infertility treatment. This must result in increased work for the HFEA.

Discrimination against the unmarried in this area may be prohibited by Article 14 of the European Convention. On the other hand, it cannot be practically or theoretically acceptable to recognize a human right in every man and woman to a baby, even by cloning, or to ART, or to successful ART, or even access to ART. A right to a baby in theoretical terms would mean a duty on some other person, or on the health service or on researchers, to make that possible. The right of a woman to a baby would have to be balanced by the right of a man to a baby, something he does not have at the moment and that would be resisted by many women because it would give a man a say in, for example, contraception and abortion. Nevertheless law reform must be in the offing.

As well as reform of the law, further light is likely to be shed on successive government attitudes to scientific advice. Reform of the HFE Act is difficult partly because of the structures erected upon it but also because any attempt to reform this area would open up the compromise reached so delicately in 1990 to scrutiny and possible destruction. It is also known that those who would restrict abortion rights would seize on any reform of the Act to insert amendments to that end. The bringing into force of regulations extending the purposes of research to include cell nuclear replacement and stem cells showed how difficult this area can be. While the government is receiving advice from the very best of scientific advisers on the various committees, it seems to be the case that a multiplicity of advice from various quarters is needed to satisfy public opinion. In relation to the recommendation that the five existing categories of research on embryos in HFE Act should be extended, it was first made in late 1998. Only after the Royal Society, the Wellcome Trust and other distinguished bodies had reported, culminating in the report by the government's Chief Medical Officer, were the proposals laid before parliament. Sometimes proposals are shelved, for example, those on regulating surrogacy; others take a long time to come to fruition, for example, donor anonymity is being further examined.

The arguments for and against donor anonymity are well known. One fact is certain: in 2007 the HFEA will have a statutory obligation to respond to children born in 1991 who will have the right, if marrying, or when reaching 18, to find out whether or not they were born by IVF and whether their fathers are anonymous sperm donors. A great deal of revision of the computer provision and the database will be needed to ensure a smooth operation. Questions such as the filter through which casual callers must pass in order to obtain this information will have to be considered. Counseling and the use to be made of the information will have to be considered. There is bound to be disappointment on the part of some children where the fathers remain unnamed and have provided very little information about themselves. The Department of Health will be conducting a consultation on whether or not the father should remain unknown; the HFEA is staying neutral in this field. If fathers are to be known, this will undoubtedly affect the operation of our database and the number of donors coming forward. If they remain anonymous, there will still be phone calls and applications to be dealt with under the provisions of the 1990 Act. Either way, another increase in work and of a sensitive nature can be foreseen.

In 2001 the HFEA may well stand at the beginning of a further leap into the future and a need for growth. The potential changes in treatment and research in embryonic stem cells are unprecedented and at the forefront of the minds of world leaders in government, religion and science. They are of major interest to the media and the public. The regulation requirements and growth in infertility and children born from licensed treatment have meant that the HFEA has had an increasingly complex and important role to play. However it has done so without any increase in its overall resource base for a number of years. It needs to change to address the government's need for a regulator that fulfils the following critical success factors: an organization which is fit for its purpose and delivers modern services; a regulator that both protects, and works with the public; a regulator that fulfils its statutory duties; a regulator that is transparent, accountable and works in partnership with key government organizations and others such as treatment and research centers, industry sources and education sources; a regulator that is responsive to emerging but critical fields in biotechnology and a regulator that communicates. There is increasing need for the robust regulation of stem cell research in embryos as a consequence of the Human Fertilisation and Embryology (Research Purposes) Regulations 2000. The changing biological context, the advance of genetics and worldwide concern regarding this issue has far reaching implications for the future of the HFEA and indeed for universities, research institutes and the biotechnological and pharmaceutical industries in Britain.

The annual funding needs to be at least doubled, and the size of the job reassessed. The clinics' contribution to the running of the HFEA should be greatly augmented. Larger premises, more staff and specialists are required, as well as the latest in IT. Members need to be well qualified and to give up more time on a voluntary basis than is recognized. The Department of Health has been slow to recognize these needs.

The history of regulation shows that work on embryos might never have been permitted in Britain at all had it not been for the existence of, at first, voluntary professional self-regulation and, subsequently, statutory controls. It has progressed with reliability and in tandem with public and peer acceptability. Judged by those criteria, regulation in principle has been a successful move. In particular issues relating to public fear, family issues, safety and widespread use, regulation has been more of a success than a failure. The security and occasional happiness offered to patients has been worth the cost of setting up the regulatory authority and has been the reward to those who have put a great deal of voluntary time and effort into doing their best for their fellow citizens. The HFEA has tried to achieve a judicial use of its powers, combined with the consultative and consensual approach to regulation. It hopes that it has gained confidence and respect from parliament, the public, patient groups and the professions. This is to the advantage of all. Parliament and Ministers, having set the framework, can rely on the Authority for the day-to-day management of the many legal and ethical issues that have emerged from the field. The HFEA in fact serves as a barrier between politicians and the public. The public, in its concern that science does not outpace ethics, is assured that breaks are in place and are functioning well. Patients are assured not only of high standards of treatment, but also being fully informed as to the benefits, risks, implications and limitations of the treatment that they are considering. And the clinics themselves enjoy the support and legitimacy of a robust, ethical system that not only controls but also protects.

This is not to say that nothing ever goes wrong. The monitoring is done with a light touch. There is no system in the world that can protect the public against total aberration or clinical or criminal behavior by someone involved in the field, but there are criminal sanctions in place and they do work.

Clearly there will be a need for international cooperation in the future. There are calls for an international treaty to ban reproductive cloning, but international lawyers will know how difficult that would be to achieve and enforce. More respect is called for, within Europe at least, for the different cultural, health and ethical standards of the different countries making up Europe and the need for one country to support another's restrictions is evident.

The future will bring further new developments which will be no less controversial than those of the recent past. IVF technologies will increasingly be employed here and in other countries for research and treatment of people with normal fertility but who risk passing on genetically inherited conditions. The HFEA has already in place a structure for managing this developing area, and, as we head further into this new territory, it is arguably more important than ever that a properly funded framework is maintained. It is equally important that this framework – the regulatory system – retains its flexibility and responsiveness. It is always necessary to seek popular support while maintaining the law and ethical standards and this is perhaps the most difficult balancing act of all.

Notes

1 Especially chlamydia with no symptoms.
2 For practical reasons a limit has been put on the number of years for which embryos and sperm may be stored.
3 Section 3.

Chapter 4

A Survey Of Regulation Of Assisted Reproductive Technology In Asia

Jothi Kumar

Introduction

Asia (excluding Australasia) is made up of about 52 countries. These countries are grouped according to their geographical location into the 4 zones of Central Asia, East Asia, South Asia and South East Asia, and also includes some Middle East countries.

Assisted Reproductive Technology (ART) was initiated in Asia in the early 1980s. In Singapore, the groundwork was undertaken in 1981, and the first pregnancy and live-birth by In Vitro Fertilization (IVF) was achieved in 1983. In Israel, the first IVF baby was born in 1982. These were the first IVF babies born in Asia. Subsequently, ART procedures were quickly adopted by a number of other countries in Asia, to treat subfertile couples. Indeed, to date, there are 25 countries in Asia where ART is clinically practiced (Table 4.1).

As with the rest of the world, ART is now widely accepted as a modality of treatment by subfertile couples in Asia. However, while its use is gaining popularity in many countries, there is generally no formal, governmental intervention to regulate and monitor the practice of ART in most of the countries in Asia.

Past surveys on regulation of ART in Asia

A study of the status of regulation of ART in Asia was first undertaken in 1990. A survey by the Committee of Reproductive Biology of the Asia-Oceania Federation of Obstetrics and Gynaecology (AOFOG) reported that the status of ART in Asia-Oceania was well established in some areas, such as IVF and GIFT, but developing in other methods, such as surrogate mother, fetal sexing and pre-embryo biopsy (Chang et al, 1994). However, laws, such as legislation, regulations and local and national Helsinki Committee Regulations (HCR), which controlled each procedure with respect to ART, were relatively less well established in most member countries, except in Australia, Japan and Singapore. In 1995, the Ethics Committee of AOFOG reported that the attitude and status of ART practice in the various countries vary in a wide range, obviously affected by local tradition, popularity in the use of such techniques, and cultural and religious concerns (Ng, T.H., 1995). Most countries did not have formal legal codes regulating the ethical aspects of ART, and each region appeared to be progressing on a "making up rules as they go along". In most countries, ART practice was regulated as part of the general laws governing the practice of medicine, rather than in the form of special legislation specifically on this subject. Schenker & Shushan (1996) noted that Taiwan was the only country (out of 20 countries in Asia with ART centers) that had specific legislation to regulate the practice of ART. In other countries, regulation was either by some kind of ministerial regulation or guidelines issued by their respective medical councils or associations. In most countries, no form of regulation was present. This wide discrepancy was attributed to the diverse cultural heritage and prevailing rules present amongst the countries in Asia, which have a major influence on the implementation of ART in Asia.

Current status of regulation of ART in Asia

An update on the state of ART regulation amongst countries in Asia where ART is practiced was undertaken in 2001. This information was collected as part of a comprehensive questionnaire which was sent to 241 clinicians/embryologists in the 25 countries in Asia where ART centers are functioning. The objective was to gather data pertaining to various aspects of the practice of ART in Asia. The countries included China, Hong Kong, Taiwan, South Korea, Japan, Vietnam, India, Sri Lanka, Bangladesh, Pakistan, Thailand, Malaysia, Singapore, Indonesia, Philippines, Israel, Egypt, Saudi Arabia, Kuwait, United Arab Emirates (UAE), Qatar, Lebanon, Turkey, Iran and Jordan. The overall response rate to the questionnaire-survey was 58%.

It is estimated that there are between 1-5 ART centers in each of the countries of Bangladesh, Pakistan, Sri Lanka, Philippines, Vietnam, Kuwait, Qatar, UAE, Iran and Jordan.

Hong Kong, Indonesia and Singapore, each have between 6-10 centers, whilst Malaysia, Saudi Arabia and Lebanon, have each between 15-20 centers. Israel, Egypt, Turkey, Taiwan and Thailand have each between 25-50 centers. It is estimated that there are more than 100 ART centers each in China, India, Japan and South Korea.

In 19 of these countries, some form of regulation is present to oversee the practice of ART. These include *legislation* in Taiwan; supervision by a *government body (Ministry of Health)* in China, Singapore, Indonesia, Vietnam and Israel; *government body (Council on Human Reproductive Technology)* in Hong Kong and *peer professional bodies* (Obstetrical & Gynaecological Societies) in Japan, South Korea and Thailand. In the Middle East countries, a *government body (Ministry of Health)* together with *official religious bodies* regulate the practice of ART in Iran, Jordan, Saudi Arabia, Kuwait, Turkey, Egypt, Lebanon, Qatar and UAE.

In the remaining six Asian countries where ART is practiced, namely Bangladesh, Pakistan, India, Sri Lanka, Malaysia and Philippines, no form of official regulation exists. This is despite the fact that in some of these countries, ART centers have been functioning since the mid-1980s.

Specific regulations

In the 19 countries where some of ART regulation exists, *official licensing of ART centers, and clinicians and embryologists* who undertake ART-related procedures, is only implemented in 11 countries. Centers are required to be licensed in Singapore, Taiwan, China, Vietnam, Indonesia, Hong Kong, South Korea, Thailand, Israel, Saudi Arabia and Kuwait, whilst clinicians and embryologists are required to be licensed only in Singapore, Taiwan and Vietnam. The validity of the licenses for the ART center, clinicians and embryologists vary between different countries. For example, in Singapore and Taiwan, the licenses need to be renewed every 2 years, whereas the license for the ART center is issued for an indefinite duration in Israel. In other countries, the validity of the licenses range between 2-3 years.

Proper record keeping and recording of all ART cases is practiced in all the countries. However, of the 19 countries with ART regulation, *compulsory reporting* of all ART treatment cycles and the outcome of treatment to the official regulatory is only required in eight countries, namely Singapore, Taiwan, Vietnam, Indonesia, Israel, South Korea, Japan and Thailand. This is generally done on an annual basis. On the other hand, *compulsory audit site-visits* to the ART center by the regulatory body is only conducted on a regular basis (1-2 yearly) in Singapore, Taiwan, South

Korea, Thailand and Indonesia. The need for ART centers to maintain a *registry of births* following ART treatment is only required in Singapore.

Reasons for lack of ART regulation

In the six countries practicing ART where no form of regulation exists, the reasons cited for this state of affairs include the following:

1. Infertility treatment and ART are not priority health issues.
2. Ignorance and lack of interest by politicians and health authorities.
3. Apathy, inertia, lack of interest and commitment by professional peer bodies.
4. Being a large and vast country, many ART centers are spread out geographically throughout the country. It is, therefore, difficult logistically to implement and audit uniform ART practice guidelines.
5. Multi-religious composition of the population in the country makes it difficult to implement and regulate uniform ART practice guidelines.
6. ART treatment is not recognized as a "subspeciality" of gynecological practice. Hence, no structured training for clinicians and embryologist exists for ART practice.
7. Most ART centers are set up in non-governmental medical centers in the private sector, without proper supervision and licensing. Governmental intervention rarely occurs in the running of ART centers in private hospitals, unless complaints arise.

While all the respondents, from the six countries without ART regulation, unanimously supported the importance and need to have some form ART "practice guidelines" and regulation to monitor the practice of ART, they indicated that they were "powerless" as a peer body to convince the relevant authorities in their respective countries to implement this. Most felt that, ultimately, public opinion and pressure will ensure that ART practice is regulated in their countries, and preferred that the regulatory body should be a "non-governmental" authority.

References

Chang, S.K., Jun, J.K., Choi, Y.M. and Moon, S.Y. (1994), 'The present status of Assisted Reproductive Technology in Asia-Oceania'. *J. Obstet. Gynaecol.*, Vol. 20, pp. 331-344.

Ng, H.T. (1995), *Report of the Chairman – Ethics Committee, Asia-Oceania Federation of Obstetrics and Gynaecology* – unpublished.

Schenker, J.G. and Shushan, A. (1996), 'Ethical and legal aspects of assisted reproduction practice in Asia'. *Human Reproduction*, Vol. 11, pp. 908-911.

Table 4.1 Asian countries with and without ART centers

With ART Centers (25)	Without ART Centers (27)
China	North Korea
Hong Kong	Laos
Taiwan	Cambodia
South Korea	Macau
Japan	Mongolia
Vietnam	Myanmar
India	Nepal
Sri Lanka	Bhutan
Bangladesh	Afghanistan
Pakistan	Maldives
Thailand	Brunei
Malaysia	Syria
Singapore	Bahrain
Indonesia	Oman
Philippines	Yemen
Israel	Iraq
Egypt	Palestine
Saudi Arabia	Armenia
Kuwait	Azerbaijan
United Arab Emirates (UAE)	Georgia
Qatar	Kazakhstan
Lebanon	Kyrgyzstan
Turkey	Nagorno-Karabakh
Iran	Russian Federation
Jordan	Tajikistan
	Turkmenistan
	Uzbekistan

Chapter 5

Canada's Proposal For Legislation On Assisted Human Reproduction

Francine Manseau

Introduction

The development and application of reproductive and genetic technologies in Canada and internationally continue to raise many profound social, ethical, legal and health issues. While some of these technologies can enhance health and well-being, others threaten human dignity and treat women, children and the reproductive process as commodities. Throughout the 1990s, many Canadians have been involved in debate and discussion around these issues, and a national consensus is emerging. There are clearly differing opinions on specific elements of these technologies but Canadians do agree that there is a need for a legislative framework to manage these technologies in a way that protects those most affected and reflects our collective values. The Government of Canada is currently considering setting in place such a legislative framework.

Historical background

As a result of the widening gap between the rapid pace of technological change and policy development needed to guide decisions about the application of reproductive technologies, the federal government created, in 1989, the Royal Commission on New Reproductive Technologies. The Final Report of the Royal Commission on New Reproductive Technologies was released in November 1993 and contained 293 recommendations. The two overarching recommendations were for federal legislation to prohibit certain practices and the establishment of a national regulatory body.

The federal government conducted extensive consultations with interested stakeholders and the Canadian public on the main recommendations of the Royal Commission. In response to the concerns of

Canadians about the application of certain problematic technologies, the Government of Canada announced, in July 1995, an interim voluntary moratorium on nine problematic applications of human reproductive and genetic technologies as the first phase in the development of an overall framework to regulate these technologies.

Bill C-47, the *Human Reproductive and Genetic Technologies Act*, was based on the moratorium and proposed to legislate the prohibition of certain applications of reproductive and genetic technologies. It was tabled in Parliament in June 1996 and received Second Reading in November 1996. At the time of tabling Bill C-47, the government also released the document *New Reproductive and Genetic Technologies: Setting Boundaries, Enhancing Health* which set out the proposed elements of a regulatory framework (prohibitions and regulations) for public discussions. However, Bill C-47 did not complete the legislative process as the result of the call of the 1997 federal election, so all efforts to legislate died at this time.

At about this time, the federal government introduced new regulations, *The Processing and Distribution of Semen for Assisted Conception Regulations*, to protect Canadian women and their children from contracting infectious diseases through donated human semen. These regulations establish uniform national standards with the objective of decreasing the risk of infectious disease transmission through semen when used in assisted conception. The regulations require appropriate procedures regarding donor screening (including testing), labeling, quarantining and record keeping prior to semen being used in assisted conception.

Before going forward with new legislative proposal, the federal government, in February 2000, undertook a final set of discussions with provinces and territories and stakeholder groups on the main parameters of a legislative and regulatory framework for assisted human reproduction. The key elements of the proposed legislation were provided in a *Workbook* which also set out a number of questions to assist participants in formulating their comments.

In May 2001, as a result of these discussions, the federal Minister of Health submitted to the House of Commons Standing Committee on Health, a committee of Parliament, draft legislation on assisted human reproduction.

The draft legislation on assisted human reproduction has two primary objectives. First, it aims to ensure that Canadians using assisted human reproduction techniques can do so without compromising their health and safety. Second, it would ensure that promising research which may help find treatments for serious conditions, such as Alzheimer's, Parkinson's and cancer, takes place within a regulated environment.

The Minister asked the Standing Committee on Health to undertake a review of the proposed legislation, including public consultations, and to provide suggestions for a possible regulatory body that would govern the implementation of the legislation. The government believed the review and consultations were needed because the draft legislation addresses health, safety, ethical, social and legal issues which are fundamental to the lives of many Canadians and deserve as broad a societal dialogue as possible.

Scope of the draft legislation

The draft legislation on assisted human reproduction is a comprehensive piece of legislation which addresses two closely related matters: the application of technology to human reproductive material (cells, gametes, etc.) for reproductive purposes; and the use of human embryos for any purpose including research and medical uses.

The proposed legislation is governed by a preamble which sets out a set of guiding principles for the legislation. It then sets out how Canada would deal with specific assisted human reproduction activities under those principles.

Certain activities would be prohibited. Activities that are considered unacceptable, such as the creation of a human clone, would be banned under statutory prohibitions. Other activities would be considered acceptable, if they were carried out in accordance with appropriate rules, called regulations. The draft legislation would give the Government of Canada the authority to develop regulations. One of the main purposes of the regulations would be to protect the health and safety of Canadians, particularly women and children.

An important set of regulations would focus on the information needs of persons using assisted human reproduction. For example, the draft legislation calls for the development of a registry to collect and keep information on sperm, ovum and embryo donations. This would make it possible for children conceived with donated gametes or embryos to learn about their medical histories.

The regulations would also define the ethical boundaries surrounding the development of the reproductive sciences that, by their very nature, tend to be at the cutting edge of applied research.

In summary, the legislation covers two types of activities: 1) those that are prohibited in all circumstances (prohibitions); and 2) those that must be carried out in accordance with the regulations (controlled activities). The prohibitions are defined in the statute and the controlled activities will be detailed at a later stage in the regulations. The draft legislation applies to two different purposes: 1) the actual assisted human reproduction activities needed to help some Canadians have children; and 2) the use of human embryos in research to help advance knowledge.

The details

Preamble

The proposed preamble to the draft legislation begins by acknowledging the benefits of assisted human reproduction technologies to individuals and to society as a whole. The preamble notes that the benefits of these technologies can be best secured by ensuring that measures are in place to protect and promote human health, safety, dignity and rights. There is an acknowledgment of the paramount need for measures to protect and promote the best interests of children who are affected by the application of these technologies.

The preamble recognizes that although all persons are affected by these technologies, women are affected more directly and significantly by their application. The principle of free and informed consent is recognized and promoted as a fundamental condition for the use of these technologies. The serious health and ethical concerns inherent in the trade of the reproductive capacities of women and men are recognized, as well as the concerns inherent in the exploitation of children, women and men for commercial purposes. Finally, the preamble recognizes the importance of preserving and protecting human individuality and the integrity of the human genome.

Prohibited activities

The draft legislation would ban certain activities, whether their purpose is to facilitate human reproduction or to advance scientific knowledge through research. The following activities would be banned:

- the intentional creation of a human clone or the intentional transplanting of such a clone into the body of a human being;
- altering the genome of a human cell or an embryo in such a way that the alteration is capable of being passed on to its descendants;

- maintaining a human embryo outside a woman's body for more than fourteen days after its creation or fertilization, excluding any time in which its development has been suspended;
- the intentional creation of an embryo solely for research purposes;
- the intentional creation of an embryo from any part of an existing embryo, fetus or intentionally transplanting such an embryo into a human being;
- transplanting any animal sperm, ovum, embryo or fetus into the body of a human being;
- the intentional use, for the purpose of creating a human being, of any reproductive material that had been transplanted into an animal;
- undertaking any activity to ensure or increase the possibility that an embryo would be of a particular sex, except for reasons related to the health of the resulting person.

Other activities would be prohibited when the purpose is for commercial gain. Without the following prohibitions, the door would be open to the commercialization of human reproduction:

- purchasing, or advertising to purchase, gametes from a donor or someone acting for the donor;
- purchasing and selling, or advertising to purchase or sell, an embryo or part of one;
- purchasing, or advertising to purchase, any other human cell or gene from a donor or someone acting for the donor with the intent of creating a human being;
- paying, or advertising to pay, a woman to become a surrogate mother;
- accepting payment for arranging the services of a surrogate mother or advertising to arrange such services;
- paying, or advertising to pay, another person to arrange for the services of a surrogate mother.

Commercial surrogacy would be prohibited to ensure that children are not viewed as commodities, and that the reproductive capacity of women are not viewed as an economic endeavor. It is important to note that although commercial surrogacy would be prohibited, surrogacy for altruistic purposes would be allowed and regulated. The prohibition is targeted at the intermediary; that is no one can pay, or accept payment, to arrange the services of a surrogate.

The focus of this prohibition is to stop individuals who are financially benefiting from commercial surrogacy arrangements. There is no intention to penalize the surrogate mother. Nor is there any intention to penalize the doctors, lawyers, or counselors who are supporting and caring for individuals involved in altruistic surrogacy arrangements. The legislation would allow reimbursement of an altruistic surrogate's reasonable expenses related to the pregnancy, as established by the regulations.

The prohibition against the commercialization of human reproductive material reflects the Canadian practice of non-commercialization of human blood, organs and tissues, and reflects the concern that a human being could result from a commercial transaction. Note that the prohibitions would criminalize the actions of everyone except for the sperm and ovum donors.

Clinics and researchers, although no longer able to purchase gametes from donors, would be free to advertise for the donation of human reproductive material and the legislation would allow for the reimbursement of a donor's expenses related to his or her donation, as established by the regulations.

Consent

The requirement for a written and informed consent is a core principle of the draft legislation. In order to respect reproductive autonomy, only individuals themselves would have the capacity to consent to the use of their gametes to facilitate human reproduction.

In order to use donated human reproductive material to create an embryo or to assist in human reproduction, the donor must have consented in writing. It would be an offence to remove human reproductive material from an individual after his or her death to facilitate human reproduction, unless the individual had provided his or her prior written consent. It would also be an offence to use an embryo in research or in the treatment of disease, injury or disability without the written consent of the donors.

Controlled activities

Activities related to human reproduction Assisted human reproduction, while beneficial, raises health and safety, as well as ethical concerns. Consequently, a license would be required to conduct controlled activities in order to protect the health and safety of individuals undergoing such procedures, particularly women as well as their offspring.

The draft legislation would require that a person obtain a license to alter, manipulate or treat any human reproductive material to create an embryo or to facilitate human reproduction, covering procedures such as donor insemination, *in vitro* fertilization and intracytoplasmic sperm injection. It would also be a requirement of the draft legislation that a person obtain a license to collect, store, transfer, destroy, import into Canada or export from Canada any sperm, ovum or embryo for reproduction purposes. These requirements would also apply when an embryo is used for research purposes.

Activities relating to research As the human embryo has the potential to develop into a human being, it deserves to be treated appropriately. Research involving a human embryo would consequently require a license to ensure that any research involving human embryos complies with the principle of respect. The draft legislation would require that a person obtain a license in order to use an embryo for the purpose of research or the prevention, diagnosis or treatment of a disease, injury or disability. The research would only be allowed on human embryos already created but not used for human reproduction, and with the written consent from the embryo providers.

Information registries A key element of the proposed legislation relates to the provision of information. It would be an offence for a licensee to accept human reproductive material from a donor, or perform any controlled activity on a person, without collecting the health information required by the regulations. This information would be required to identify concerns about the health and safety of donated reproductive material, and the health and safety and efficacy of any assisted reproduction procedure. Prior to the collection of the necessary health information, the licensee would have to provide the donor or the patient with written information about the requirements of the legislation respecting the care and control of such information.

The draft legislation would require that a donor/offspring information registry be maintained. The registry would contain information relating to donors, persons undergoing an assisted human reproduction procedure using the donor's gametes or embryos, and persons conceived through the use of these procedures. The legislation would allow the disclosure of non-identifying information about a donor to a person undergoing an assisted human reproduction procedure with the donor's gametes or embryo, or to the person conceived with those gametes or embryos and his or her descendants. Identifying information about the donor could only be released with the donor's written consent.

There is also a requirement in the draft legislation to maintain a public information registry available for public scrutiny. The registry would contain information relating to the Minister's administrative powers under the draft legislation, listing of licensed infertility treatment and research centers. Also, as part of the public information registry, the Minister would be required to produce, on an annual basis, a report on the outcomes of infertility treatment procedures performed by all licensed facilities.

Enforcement

The proposed legislation would include inspection and enforcement provisions similar to those found in other federal health legislation. These provisions include the powers to enter premises and conveyances, examine, seize and analyze material and records, provide directions to the licensee respecting the provision of the controlled activity, provide for the disposal of material, and operate the facility if need be.

Offenses and punishment

Persons seeking to provide a controlled activity would have to obtain a license. Licenses would include terms and conditions pertaining to the undertaking of a controlled activity. Licenses could be amended, renewed, suspended, or revoked.

A breach of the terms of a license could lead to its suspension, revocation, or renewal with terms and conditions. In addition, it could lead to a charge under the offence provision of the legislation and attract imprisonment and/or a fine.

Under the proposed legislation, a person convicted of contravening any of the prohibitions would be liable on indictment to a fine of up to a maximum of $500,000 Canadian dollars or imprisonment for a maximum of ten years or both. If, however, the individual is convicted of a summary offence, they would be liable to a fine of up to a maximum of $250,000 Canadian dollars or imprisonment for a maximum of four years or both.

A person who contravenes any of the controlled activities and the privacy provisions, or any terms and conditions of a license (with significant harm), would be liable on conviction on indictment to a fine of up to a maximum of $250,000 Canadian dollars or imprisonment for a maximum of five years or both. If a person is found guilty on a summary conviction, they are liable to a fine of up to a maximum of $100,000 or imprisonment for a maximum of two years or both.

The regulatory body

The House of Commons Standing Committee on Health was asked, in their review of the draft legislation, to make recommendations on the type of regulatory body that should be created to oversee the implementation of the proposed legislation.

As drafted, the proposed legislation would allow the Minister of Health to create a regulatory body internal to the federal Department of Health. While an external regulatory body is also an option that has been identified, the provisions for such an external body have not been included in the current draft legislation pending the Standing Committee's recommendations on this matter. If the decision is taken to create an external body, legislation would need to be drafted accordingly.

Previous consultations undertaken by Health Canada with external stakeholders and interest groups, indicates that a general consensus for an external regulatory body exists. Such a body should be broadly representative of all parties interested in assisted human reproduction, transparent in its processes, be accountable and it should report to the Minister of Health.

Four basic functions are being considered for the regulatory body, including:

- *Policy development* To provide advice to the Minister on assisted human reproduction issues, such as recommending changes to the regulations; to follow assisted human reproduction developments in Canada and around the world.
- *Licensing and enforcement* To issue licenses to public for private organizations and persons conducting controlled assisted human reproduction practices, services or research. Inspections would be used as a means to ensure compliance by licensees with terms and conditions for controlled activities.
- *Health information* To maintain health information registries on donors, recipients, persons born of assisted human reproduction and to report on the outcomes of treatments and procedures and research.
- *Communications and consultations* To conduct ongoing consultations and communications with stakeholders, interest

groups and the public to resolve issues or concerns and to provide information to prospective clients on their assisted human reproduction choices and to the Canadian public on related human reproduction matters.

Conclusion

The House of Commons Standing Committee on Health released, in December 2001, its report and recommendations in response to the Government of Canada's draft legislation on assisted human reproduction. The government is now paused to review the draft legislation taking into account the work of the Standing Committee and will be presenting a revised legislative framework as quickly as possible.

The Government of Canada is aware of the high degree of interest and desire to have legislation on assisted human reproduction in place, especially given recent developments in the area of cloning. It is committed to presenting a legislative framework that is comprehensive, dynamic and indeed a "made in Canada" response to this complex, and at times controversial, set of issues.

Chapter 6

Regulating ART In The USA: A Mixed Approach

Jennifer Gunning

The approach to the regulation of assisted reproductive technologies (ART) in the United States is a mixture of restriction and laissez faire. Human embryo research is severely restricted at the federal level but the provision of clinical services is largely subject to professional self-regulation. Although the Fertility Clinic Success Rate and Certification Act of 1992 established a federal monitoring mechanism through the Centers for Disease Control (CDC), health care is a States' right. It is up to individual States to decide whether to implement the Act and federal sanctions are not possible.

Early debate

Discussion of the ethical aspects of ART has been active in the USA since the early days of in vitro fertilization and well before the birth of the first IVF baby. The creation of the first human embryo in vitro was reported in 1969 (Edwards, Bavister and Steptoe, 1969). While there was little or no ethical debate in the United Kingdom, where the research had been carried out, the issues were discussed at the twelfth meeting of the panel on science and technology of the US House of Representatives Committee on Science and Aeronautics in January 1971. There James Watson - the Nobel Laureate who solved the structure of DNA with Francis Crick - presented a paper entitled 'Potential consequences of experimentation with human eggs' (Watson, 1971). His view of the possible consequences was not sanguine. He said:

> Some very hard decisions may soon be upon us. For it is not obvious that the vague potential of abhorrent misuse should weigh more strongly than the unhappiness which thousands of married couples feel when they are unable to have their own children. Different societies are likely to view the matter

> differently and it would be surprising if all come to the same conclusion. We must, therefore, assume that techniques for the in vitro manipulation of human eggs are likely to be general medical practice, capable of routine performance in many major nations within some 10 to 20 years.

Aware of the work by John Gurdon in Oxford on clonal reproduction in frogs, Watson believed that action needed to be taken to prevent human cloning experiments and to consider the consequences of further advances in human embryology. He said:

> This is a matter far too important to be left solely in the hands of the scientific and medical communities. The belief that surrogate mothers and clonal babies are inevitable because science always moves forward ... represents a form of laissez faire nonsense dismally reminiscent of the creed that American business, if left to itself, will solve everybody's problems. Just as the success of a corporate body in making money need not set the human condition ahead, neither does every scientific advance automatically make our lives more meaningful.

He believed that international agreement was the preferred method of control.

> I would thus hope that over the next decade wide reaching discussion occurs, at the informal as well as the formal legislative level, about the manyfold problems which are bound to arise if test tube conception becomes a common occurrence. On some matters a sufficient international consciousness might be apparent to make possible some forms of international agreement before the cat is totally out of the bag. A blanket declaration of the worldwide illegality of human cloning might be one result of a serious effort to ask the world in which direction it wishes to move.

Twenty years on such broad based international agreement has yet to be achieved although a European Convention came into force in December 1999, having been ratified by six countries.

Watson was speaking at the start of a debate that gathered pace soon after and has continued ever since. Another contributor to the early debate was Leon R. Kass, from the Committee on the Life Sciences and Social Policy at the US National Research Council, now chairman of The President's Council on Bioethics. He was also concerned about the potential dangers to the resultant offspring and wrote in the *New England Journal of Medicine* (Kass, 1972):

> Because the new procedures for in vitro fertilization and laboratory culture of human embryos probably carry a serious risk of damage to any child so generated, there appears to be no ethical way to proceed. One cannot ethically choose for a child the unknown hazards that he must face and simultaneously choose to give him life in which to face them.

Kass suggested that a profession-wide, self-imposed moratorium should be implemented to allow interdisciplinary discussion internationally and in and out of government.

Regulating human embryo research

To understand the current position in the USA in relation to the funding and regulation of human embryo research one needs to consider the historical context. In 1973 the Department of Health Education and Welfare (now the Department of Health and Human Services) began to discuss and to draft regulations relating to in vitro fertilization and embryo transfer. Finally, in 1975, these regulations were published (45CFR46.204(D)) stating that:

> No application or proposal involving human in vitro fertilization may be funded by the Department or any component thereof until the application or proposal has been reviewed by the Ethics Advisory Board and the Board has rendered advice as to its acceptability from an ethical standpoint (US Federal Register, 1975).

This meant that, without approval from the Ethics Advisory Board, no federal funds were available for research and, as no Ethics Advisory Board was appointed no research was undertaken.

In 1977 the Department of Health Education and Welfare (HEW) received an application for support for research involving human in vitro fertilization and in September that year and Ethics Advisory Board was eventually appointed. Its Chairman was James Gaither, an eminent San Francisco lawyer and Vice Chairman David Hamburg, President of the Institute of Medicine at the National Academy of Sciences. Meeting in May 1978, the Board agreed to review the research proposal. However, following Edwards and Steptoe's success in the field a couple of months later, the Board was asked by the Secretary of the HEW to broaden its remit to consider the scientific, ethical, legal and social issues surrounding human in vitro fertilization and embryo transfer in general. Over the next nine months the Board held a series of eleven public hearings throughout the United States, listening to the views of private individuals, professional societies and public interest groups. Scholars and experts were approached in the fields of reproductive science, ethics, theology, law and the social sciences to prepare reports and to give evidence to the Board. Overall the Board received over 2,000 pieces of correspondence. Its report was issued in May 1979 (Ethics Advisory Board, Department of Health Education and Welfare, 1979).

The Board came to five conclusions. The first was that 'The Department should consider support of carefully designed research involving in vitro fertilization and embryo transfer in animals, including non-human primates. The Board had received conflicting evidence as to the conclusiveness of the research so far undertaken in animals in assessing the risks associated with IVF and embryo transfer.

The second conclusion of the Board was that it found it acceptable from an ethical standpoint to undertake research involving human in vitro fertilization and embryo transfer with the following provisions.

A. If the research involves human fertilization and embryo transfer, the following conditions are satisfied:

1. the research complies with all appropriate provisions of the regulations governing research with human subjects (45CFR46);
2. the research is designed primarily: (A) to establish the safety and efficacy of embryo transfer and (B) to obtain important scientific information to that end not reasonably obtainable by other means;
3. human gametes used in such research will be obtained exclusively from persons who have been informed of the nature and purpose of the research in which such materials will be used and have specifically consented to such use;
4. no embryo will be sustained in vitro beyond the stage normally associated with the completion of implantation (14 days after fertilization) and;
5. all interested parties and the general public will be advised if evidence begins to show that the procedure entails risks of abnormal offspring higher than those associated with natural human reproduction.

B. In addition, if the research involves embryo transfer following human in vitro fertilization, embryo transfer will be attempted with gametes obtained from lawfully married couples.

The Board carefully stated that this conclusion did not seek to address whether the Department should or should not support such research but to provide general guidance to Institutional Review Boards or other groups who might be asked to review this sort of research and who might benefit from the Board's deliberations. It considered that much remained to be learned about the safety and effectiveness of IVF and embryo transfer before they could be considered standard, accepted medical practice and that research designed to improve the safety and efficacy of these procedures would be ethically acceptable.

The Board also discussed research involving in vitro fertilization leading to the general advancement of knowledge. Provided that such research was performed as a corollary to research undertaken primarily to establish the safety and efficacy of IVF and embryo transfer, the Board agreed that this would be ethically acceptable.

The Board's third conclusion was that, whilst it considered it ethically acceptable to the Department to support research involving human IVF end embryo transfer, it decided not to address the level of funding, if any, which such research should be given. That is, the Board did not wish to get involved in any political decision making as to the priority such research should be given in the allocation of public research funds. This was a matter for the Department itself.

The fourth conclusion of the Board was that The National Institute of Child Health and Human Development (NICHD) and other appropriate agencies should work with professional societies, foreign governments and international organizations to collect, analyze and disseminate information derived from research (in both animals and humans) and clinical experience throughout the world involving in vitro fertilization and embryo transfer. This derived from the foresight of the Board that IVF techniques would be rapidly taken up and adopted world wide and that sometimes procedures would be performed without the benefit of research design or experimental controls. The Board considered that it would be unfortunate not to have access to the results of such research notwithstanding the fact that well-designed investigations were preferable. The Board was anxious that practitioners of IVF should not be ignorant of the findings of others.

Finally, the Board concluded that The Secretary of State should encourage the development of a uniform model or law to clarify the legal status of children born as a result of IVF and that funds should be made available for the development of such legislation.

Considerable resources were put into the deliberations of the Ethics Advisory Board in terms of expense as well as the time and energy of its members. Despite this the Secretary of State took no action on its report. The Board's charter expired in 1980 and it was not reinstated. In 1980 Congress created the President's Commission for the Study of Ethical Problems in Medicine and Biomedical and Behavioral Research. This endorsed the Board's conclusions but carried out no further analysis. As there was no EAB researchers were discouraged from seeking NIH funds until 1988 when US Office of Technology Assessment published a report, *Infertility: Social and Medical Choices*, which was highly critical of the DHHS for its posture (US Congress, Office of Technology Assessment, 1988). Subsequently, the NIH received a research proposal it wished to fund. Although, in June 1988, an announcement was made by the Assistant Secretary for Health that the EAB would be re-established, and a proposed charter later issued for comment, no action was taken. Federal funds remain unavailable for research involving human in vitro fertilization and this position was further formalized in law by Congress in 1996 (Pub. L. No.144-99, Title I, Para 128, 110 Stat. 26,34 (1996)). The law prohibits the use of federal funds 'for the creation of a human embryo or human embryos for research purposes or research in which a human embryo or embryos are destroyed, discarded or knowingly subjected to risk of injury or death'. This does not, of course, mean that no human embryo research is undertaken in the USA. Private clinics and company laboratories can and do undertake such research without restriction and the 1988 applicant to the NIH obtained support from a private foundation and completed his research without federal funds.

At the State level there has been some legislative activity with regard to human embryo research. For example, among the first to legislate were Illinois and Louisiana which passed statutes prohibiting human embryo research (Ill.Ann.Stat.ch.38. paras 88-26,32 (1987) and Ill.Ann.Stat.ch. 38. para 9-1.2(3)(b) (1989); La.Rev.Stat.Ann.Secs.14:87, 121-133 (1987)). Pennsylvania, on the other hand, permitted human embryo research provided that reports, including details of the research, were filed and available for public inspection. Failure to file such a report would incur a fine of $50 per day (Pa.Stat.Ann.tit.18, Secs. 3203, 3216 (1983)).

Regulating clinical services

The provision of ART in the United States is very diverse with facilities ranging from large university medical complexes to single doctor offices and the professional qualifications of those offering ART procedures are equally disparate. The costs of ART are largely met out of insurance and some states, for example, Delaware, Maryland, Hawaii, Texas, Arkansas and Massachusetts, have mandated that IVF be covered by insurance. Cover under state Medicaid programs varies; many will pay for drugs, counseling and surgical procedures related to infertility treatment, some including ART.

Regulation of ART has been almost entirely down to professional self-regulation, first through the American Fertility Society (now the American Society for Reproductive Medicine (ASRM)) and more recently in conjunction with the Society for Assisted Reproductive Technology (SART). The ASRM provides guidelines for the provision of infertility services, most recently in 1996 (ASRM, 1996), and its Practice Committee produces regular reports and guidelines on aspects of ART. In 1999 guidelines were issued for advertising by ART programs (ASRM, 1999(a)). This required clinics to comply with the guidelines of the Federal Trade Commission, not to mislead patients or the public in the way in which data are presented and to make clear the method used to calculate success rates. Advertising clinics quoting program statistics are required to include the statement 'A comparison of clinic success rates may not be meaningful because patient medical characteristics and treatment approaches may vary from clinic to clinic'.

Also in 1999, the ASRM issued guidelines on the number of embryos transferred (ASRM, 1999 (b)). These guidelines suggest that the number of embryos transferred should be agreed upon by the physician and the patients and informed consent obtained. While it is suggested that in patients with the most favorable prognosis usually no more than two good embryos should be transferred, this may be increased depending on the case. In patients with above average prognosis three embryos may be transferred, in patients with an average prognosis no more than four embryos and in patients with below average prognosis up to five embryos may be transferred. In donor egg cycles it is recommended that the age of the donor should determine the number of embryos transferred. These guidelines are much more flexible than those used in the UK, for instance.

Most recently the ASRM, together with the SART, has addressed preimplantation genetic diagnosis (ASRM and SART, 2001). This report summarizes the clinical methodology and experience of this technology and cautiously recommends its use. All ASRM guidelines and reports are accessible at www.asrm.org.

Against this background of flexible professional self-regulation there has been federal activity addressing the regulation of ART services. In April 1992 a bill was introduced into Congress to provide for reporting pregnancy success rates in ART programs and for the certification of embryo laboratories. After discussion and amendment in Congress and Senate, the Fertility Clinic Success Rate and Certification Act (FCSRCA; Pub. L. 102-493,42 U.S.C. 263a-1 et seq.) was signed in October 1992 by the former President Bush. The law was intended to provide the public with reliable, standardized information about the effectiveness of infertility services and to assure the quality of such services. The principal provisions of the law address ART programs and the evaluation of their services and success rates; the certification of embryo laboratories; the accreditation of organizations; the processes of certification, revocation and suspension; and the publication of success rates. It was required that within two years of the enactment of the law each ART program should submit an annual report to the Centers for Disease Control (CDC) containing its pregnancy success rates, the identity of each embryo laboratory used by the program and whether the laboratory was certified. Assisted Reproductive Technology is defined as 'all treatments or procedures which include the handling of human oocytes or embryos including in vitro fertilization, gamete intra fallopian transfer, zygote intra fallopian transfer and other such specific technologies as the Secretary may include in this definition, after making public any proposed definition in such manner as to facilitate comment from any person including any Federal or other public agency'. An embryology laboratory is defined as 'a facility in which human oocytes are subject to assisted reproductive technology treatment or procedures based on the manipulation of oocytes or embryos which are subject to implantation'.

The SART has been collecting and publishing data on ART success rates since 1989. Since 1995 SART has worked together with the CDC to gather the data required under the Act and to issue reports. The latest data available are those for 1999 (www.cdc.gov/nccdphp/drh/ART99). Of the 399 clinics offering ART services 370 (95%) submitted data. The number of ART cycles reported was 86,822 resulting in 21,501 live birth deliveries and 30,285 live babies born. More than 75% of these cycles used fresh, non-donor eggs or embryos. Donor eggs were offered by 84% of the reporting clinics and donor embryos by 51%. Single women were treated by 83% of clinics and gestational carriers (surrogates) were offered by 61%. Only 79% of the reporting clinics had verified laboratory accreditation.

The FCSRCA requires that the Secretary of State for Health and Human Services, through the CDC, develop a model program for the certification of embryo laboratories, to be carried out voluntarily by interested States. In 1998 the CDC published a proposed model for the certification program and invited comments (Federal Register, 1998). The proposed model program was developed in consultation with the ASRM and SART, together with the College of American Pathologists and the voluntary Reproductive Laboratory Accreditation Program which currently provides oversight of over one third of the US embryo laboratories affiliated with ART programs. Standards and guidelines from other professional organizations, State, Federal and international programs were also used as resources. Laboratories were visited and work sessions were held with technical consultants. On the publication of the proposed model program a 60-day comment period was allowed for the public to submit their concerns and recommendations.

In response to this public consultation a total of 15 letters was received. Four were from professional organizations, one from a consumer advocacy group, one from a manufacturer and the rest from individuals employed in embryo laboratories or affiliated with infertility clinics. The letters contained approximately 60 comments and six of the respondents were complimentary of the CDC's efforts to develop the model program. In July 1999, the CDC published the results of its consultation together with the revised Model Certification Program for Embryo Laboratories (Federal Register, 1999). The Model Certification Program (MCP) will be distributed to State officials and health authorities to encourage their assistance in the State adopting the program. But the model program does not provide a federal oversight role. Health care is a State issue and the MCP has been developed as a resource that can be taken up by States in developing their own programs for regulating embryo laboratory standards. Federal pressure can only be brought to bear through the publication of the certification/accreditation status of embryo laboratories affiliated with ART programs or clinics as part of the annual CDC reports on ART success rates.

Cloning and stem cell research

In February 1997 the announcement was made that Scottish scientists had succeeded in cloning a sheep using the nucleus of a mammary cell and an enucleated egg. This was the first time that a fully developed animal had been produced using cell nuclear transfer. Almost immediately President Clinton declared a ban on federal funding for research aimed at cloning human beings. The recently appointed National Bioethics Advisory Commission (NBAC) was asked to consider the ethical and legal issues surrounding cloning and to report within 90 days. The NBAC report was sent to the President on 6 June 1997 (National Bioethics Advisory

Commission, 1997). The report deals with the science involved in cell nuclear transfer, the positions of the major religious philosophies on human cloning, and the ethical, legal and constitutional issues involved. The report concludes that, because of serious ethical and safety concerns, it is currently morally unacceptable for anyone in the public or private sector to attempt to create a human individual through cell nuclear transfer and that the current federal moratorium should continue. The private sector should be asked to comply voluntarily. Federal legislation should be enacted to ensure this but the legislation should contain a sunset clause to review whether the prohibition should be continued. The Commission also recommended that the wide ongoing deliberation should continue.

In 1998 two papers were published reporting the successful derivation of human embryonic stem cell lines (Thomson et al, 1998; Shamblott et al, 1998). Both the NIH and NBAC went forward simultaneously to develop guidelines for research involving human embryonic stem cells. Both reported in 1999 (National Institutes of Health, 1999; National Bioethics Advisory Commission, 1999). NIH was keen to begin funding work on embryonic stem cells but its recommendations had to fall within the scope of permissible federal funding. It concluded, in accordance with NIH legal counsel's ruling, that federal funds could be used to support this work as long as the stem cells were not derived from the embryos by a researcher using federal funds. Since deriving stem cells required the destruction of human embryos federal funding could not be used for that aspect of the research. The NBAC also concluded that human embryonic stem cell research should be eligible for federal funding but this should also include the derivation of these cells. The Commission saw no meaningful ethical distinction between derivation and use of embryonic stem cells. These two reports stirred up the political debate but no action was taken until the change of administration.

On August 9 2001 President George W. Bush broadcast some remarks on stem cell research to the nation. In these remarks he said

> As a result of private research, more than 60 genetically diverse stem cell lines already exist. They were created from embryos that have already been destroyed, and they have the ability to regenerate themselves indefinitely, creating ongoing opportunities for research. I have concluded that we should allow federal funds to be used for research on these existing stem cell lines, where the life and death decision has already been made.

He continued

> This allows us to explore the promise and potential of stem cell research without crossing a fundamental moral line, by providing taxpayer funding that would sanction or encourage further destruction of human embryos that have at least the potential for life.

He announced that in 2001 the government would also spend $250 million on research on umbilical cord, placenta, adult and animal stem cells which do not incur the same moral dilemma and which, by implication, might provide alternatives. President Bush also announced the establishment of a President's Council on Bioethics 'to monitor stem cell research and to recommend appropriate guidelines and regulations, and to consider all of the medical and ethical ramifications of biomedical innovation'. The Council would be chaired by Leon R. Kass, an early contributor to the human embryo research debate, and would first meet in January 2002. The remit of the NBAC expired in October 2001 and the Commission was effectively replaced by the new President's Council.

More recently President Bush has become more outspoken in his opposition to any form of cloning. At the time of writing, two bills are before the US Senate. One would ban only human reproductive cloning, the other would ban all forms of cloning including therapeutic cloning. George Bush has spoken out urging the Senate to ban all research on human cloning. He has said that anything short of a full ban would be unethical and nearly impossible to enforce.

References

ASRM, (1996), Practice Committee, *Guidelines for the Provision of Infertility Services*, ASRM, Birmingham, Alabama, USA.

ASRM, (1999a), Practice Committee, *Guidelines for Advertising by ART Programs*, ASRM, Birmingham, Alabama, USA.

ASRM, (1999b), Practice Committee, *Guidelines on Number of Embryos Transferred*, ASRM, Birmingham, Alabama, USA.

ASRM and SART, (2001), *A Practice Committee Report: Preimplantation Genetic Diagnosis*, ASRM, Birmingham, Alabama, USA.

Edwards, R.G., Bavister, B.D., and Steptoe, P.C. (1969), 'Early stages of fertilization in vitro of human oocytes matured in vitro', *Nature*, Vol. 221, pp. 632-635.

Ethics Advisory Board, Department of Health Education and Welfare (1979), *Report and Conclusions: HEW Support of Research involving Human In Vitro Fertilization and Embryo Transfer*, Appendix 12, May 4.

Federal Register Notice, November 6, 1998 (63 FR 60178), *Implementation of the Fertility Clinic Success Rate and Certification Act of 1992: Proposed Model Program for the Certification of Embryo Laboratories.*

Federal Register Notice, July 21 1999, Vol. 64, No. 139, *Implementation of the Fertility Clinic Success Rate and Certification Act of 1992: A Model Program for the Certification of Embryo Laboratories.*

Kass, L.R. (1971), 'Babies by means of in vitro fertilization: unethical experiments on the unborn?', *New England Journal of Medicine*, Vol. 285, No. 21, pp. 1174-1179.

National Bioethics Advisory Commission, (1997), *Human Cloning*, Rockville, Maryland, USA.

National Bioethics Advisory Commission, (1999), *Ethical Issues in Human Stem Cell Research*, Rockville, Maryland, USA.

National Institutes of Health, (1999), *Draft National Institutes of Health Guidelines for Research Involving Human Pluripotent Stem Cells*, Bethesda, Maryland, USA.

Shamblott, J.J., Axelman, J., Wang, S., Bugg, E.M., Littlefield, J.W., Donovan, P.J., Blumenthal, P.D., Huggins, G.R. and Gearhart, J.D. (1998), 'Derivation of Pluripotent Stem Cells from Cultured Human Primordial Germ Cells'. *Proceedings of the National Academy of Scences USA*, Vol. 95, pp. 13726-13731.

Thomson, J.A., Itskovitz-Eldor, J., Shapiro, S.S., Waknitz, M.A., Swiergiel, J.J., Marshall, V.S. and Jones, J.M. (1998), 'Embryonic Stem Cell Lines Derived from Human Blastocysts'. *Science*, Vol. 282, pp. 1145-1147.

US Congress, Office of Technology Assessment (1988), *Infertility: Medical and Social Choices*, OTA-BA-358, US Government Printing Office, Washington D.C., USA

US Federal Register (1975), Vol. 40 (154), August 8, p. 33529.

Watson, J.D. (1971), 'Potential consequences of experimentation with human eggs'. Presented at the twelfth meeting of the panel on science and technology, Committee on Science and Astronautics, United States House of Representatives, Washington D.C., USA, January 28.

Chapter 7

Development Of Regulation Of Assisted Reproduction: A World View Of Early Days

Louis Waller

I

Weltanschauung - The German word, meaning literally a 'world view', has marched into English, but is used to mean a person's whole conception of life, or of some important part of it. This is my world view of our subject. Since there are distinguished participants in this International Symposium who will contribute their assessments of ART regulation in the United Kingdom, in New Zealand, in Asia, in Canada, in the USA, and in Europe, I shall confine myself to early days in Australia in general, and in Victoria in particular.

II

In the decade following the birth of Candice Reed, in Melbourne on 23 June 1980, the governments of each Australian state instituted an inquiry into the new phenomenon - in vitro fertilization, and its attendant developments.[1] The most substantial and most protracted inquiry was undertaken in New South Wales. An Advisory Committee on Human Artificial Insemination was established by the Minister for Youth and Community Services in March 1982. That was subsumed into a reference on artificial conception given by the Attorney-General to the New South Wales Law Reform Commission in October 1983. The Commission published its first Discussion Paper on Donor Insemination late in 1984, and two further Discussion Papers, on IVF and Surrogate Motherhood. Three Reports, on Human Artificial Insemination, on In Vitro Fertilization, and on Surrogate Motherhood, appeared in June 1986, July 1988, and December 1988. Nine years elapsed before the New South Wales Health

Department published an excellent Issues Paper to canvass opinion and invite comments on proposals for ART legislation in that State.[2]

The inquiry reports in South Australia and Western Australia engendered legislative responses. In South Australia the *Reproductive Technology Act* was enacted in 1988. The Western Australian Parliament passed the *Human Reproductive Technology Act* in 1991. Specific legislation on surrogate motherhood was enacted in Queensland, in Tasmania, in the Australian Capital Territory, and in South Australia.[3] These enactments call to mind the *Surrogacy Arrangements Act 1985*, passed by the United Kingdom Parliament in the wake of the notorious *Baby Cotton* case.

The history of regulation in Victoria is of particular significance for several reasons. Before turning to it, however, I recall that all these IVF inquiries were preceded by the activities of the Standing Committee of Attorneys-General, aimed at the production of uniform legislation on status issues arising from donor insemination. It should not be forgotten that many children were born from donor insemination in post World War Two Australia before that fundamental question of the legal relation between the husband of the donee mother and her child was first considered. Status legislation has been passed in all Australian States and Territories, though forms and details vary. In essence, these statutes create the paternal relationship between the child born through ART using donor sperm, and the partner of her or his mother, or the maternal relationship between such a child and the woman who bore her or him as a result of the gift of an oocyte.[4]

III

The Thompson Liberal government established the Committee to Consider the Social, Ethical and Legal Issues Arising from In Vitro Fertilization in March 1982. The Cain Labor government assumed office the following month, and completed the work of its predecessor by appointing the members of the Committee, which began its work in May 1982.[5] The questions and the criticisms which were everywhere expressed led to this response. The Anglican Primate of Australia and the Catholic Archbishop of Melbourne published statements on their general and wide-ranging concerns about the "test-tube baby program" in January 1982. Other members of the Victorian community asked: "How would IVF children cope with the knowledge of how they were conceived?". Professor - now Emeritus Professor - Carl Wood, the acclaimed pioneer in the use of IVF in the treatment of infertile couples in Victoria, called, not for the first time, for the establishment of a government inquiry into "test-tube ethics".[6]

The Committee's terms of reference were broad. One of them required the Committee to deliver an Interim Report no more than three months after beginning its work. Submissions were widely solicited, a public hearing was held, and expert opinions were presented. That first Report, published in September 1982, contained several fundamental, unanimous recommendations. The first was that IVF in the most common situation, where no donor gametes were used and the embryos successfully formed were transferred to the wife, was acceptable in Victoria. The second was that comprehensive information and specialist counseling must be provided for each infertile couple, who would thereupon consent to proceed knowing the nature and the quality of the procedures, the risks attendant on them, and the disappointments which might often ensue. The third, and perhaps the most significant, was that IVF programs should be regulated by legislation. This recommendation was not supported by, nor was it the result of a lengthy examination of other modes of regulation. The Committee knew that the National Health and Medical Research Council (NHMRC), the principal Australian advisory and funding body in the field, had established a Medical Research Ethics Committee, under the chairmanship of the late Emeritus Professor Richard Lovell, which was formulating guidelines on the ethical aspects of research and other matters related to the use of ART. The result was what became Supplementary Note 4 to the NHMRC Statement on Human Experimentation, issued in October 1982. Those guidelines were addressed to institutional ethics committees, themselves established pursuant to NHMRC recommendations.

I have, in other places, offered an explanation for the IVF Committee's unequivocal recommendation for statutory regulation.[7] It begins with the assessment of Sir Ninian Stephen, when a Justice of the High Court of Australia. Although his purpose was to contrast the law made by courts with the law made by parliaments, his statement applies generally to statutes. He said this:

> The passage of a measure through the legislature confers a unique stamp of democratic legitimacy, valuable in a country possessing democratic traditions. Moreover, the legislative process is exposed to, and provides a safety value for, those community pressures which, if not released in this way, may otherwise build up to levels dangerous to the system itself. An elected legislature as the identified and visible maker of laws can be seen to be responsive to legitimate pressures and to the strongly held views of the community.[8]

What more effective way is there, in our modern democratic society, of enunciating those values which society considers of high importance? Some critics of legislation to regulate ART contend that Parliament should not interfere in the private lives of adult competent members of our community. It is the argument of autonomy, and it is, in general, an argument which commands respect. But this is not the general case. What may, for example, be said about anti-contraception legislation, where an argument about the law staying outside the matrimonial bedroom had and has a powerful impact, cannot be sensibly propounded in the case of assisted reproductive technologies and particularly where the formation of embryos is concerned. As a community, we all have a powerful interest in what is done in public and private hospitals, however denominated, and even in doctors' surgeries and clinics, where wide variations in genetic parenting may be initiated. As a community, we have an interest in developments in what Australia's *Family Law Act 1975* entitles "the natural and fundamental group-unit of society".[9] That unit is the family.

IV

In August 1983 the IVF Committee published its Report on the use of Donor Gametes in IVF. Its most significant recommendations were those on recording and preservation of comprehensive information on donors where a child was born as a consequence, and on access to those records.[10] The third and final Report, on embryo formation, experimentation, storage and surrogate motherhood in the context of IVF, was published in August 1984. Its principal recommendations were

- that embryos shall not be formed for the purpose of destructive research but;
- that embryos formed for the purpose of an ART program, but not transferred - and perhaps not transferable - may be donated for approved research[11] and;
- that commercial surrogacy arrangements and associated advertising should be prohibited.[12]

The response of the government, and the Parliament, to these three Reports followed their wide exposure to the Victorian community, and the consideration of submissions which resulted from that process. Before the final Report was published, the Cain Labor government stated its intention to introduce legislation to give effect to the Committee's recommendations.

In the autumn session of 1984, the *Status of Children (Amendment) Act* was passed, to which I alluded earlier. It dealt with all status issues arising from donor births. The *Infertility (Medical Procedures) Act* was passed in November 1984. This pioneering statute provided for the regulation of ART programs by ministerial approvals, and

- the provision of information to infertile couples considering ART treatment procedures;
- the mandatory counseling of those couples, and of donors;[13]
- the transmission of comprehensive information about donors where a live birth ensued, and its preservation in a Central Register;[14]
- the prohibition of the formation of embryos other than for approved ART procedures;[15]
- the prohibition of cloning (though the expression was left undefined);[16]
- the prohibition of commercial surrogacy arrangements, and of advertisements related to them;[17]
- the establishment of the Standing Review and Advisory Committee on Infertility, responsible for considering proposals for permissible embryo experimentation and related issues, and for advising the Minister for Health in relation to infertility and procedures for its alleviation.[18]

The enactment of this statute was a milestone on the road of regulation. It was proclaimed in stages, and came into full force and effect on 1 July 1988. Before that date it was amended, again with bipartisan support, to encompass GIFT procedures and to provide for permissible experiments on fertilized oocytes before that stage of development entitled 'syngamy'.[19] The history of the first Act of its kind, in Australia and the common law world, has been described and considered by me and other writers. The Act attracted general and specific criticisms.[20] Cecil Fifoot, a perceptive and subtle legal historian, described a statute as "the most ambitious of documents".[21] But he continued that it "is also the most equivocal". The regulation of human behavior by the use of words inevitably evokes problems and controversies about the core and penumbral meanings of those words. Despite these responses, the Act worked effectively in the main, and paved that same road for the passage of the more comprehensive *Infertility Treatment Act* 1995, passed, once more, with bipartisan support.

It would be an overestimation to describe the Victorian statute as a model for others. The South Australian and Western Australian enactments are cast in another mould.[22] But in my opinion the Act of 1984 served to stimulate and encourage those States to pursue regulation by legislation. It is more difficult to establish that connection with the United Kingdom's *Human Fertilisation and Embyrology Act 1990*, though there are some significant similarities, and equally significant differences. The proposed Canadian legislation may also engender comparisons, but they will be with the current Victorian statute.

V

The anguish of infertility in the lives of married and de facto couples stimulated the development of in vitro fertilization. Concerns, and indeed fears, about this phenomenon led to the inquiries already described. Today, almost a quarter of a century has elapsed since Louise Brown was born. Still issues related to ART abound.[23] This is no cause for surprise. Infertility and attempts at its alleviation, or perhaps better its circumvention, is a condition long recognized in our cultural heritage.[24] Effective legislation, however, is less than 20 years old. It will be a matter for later judgment to determine the worth and value of those measures. I hope the result of that will be a fair and favorable one.

Notes

1 Louis Waller (1992), "Australia: The Law and Infertility – the Victorian Experience", in Sheila A M McLean ed *Law Reform and Human Reproduction*, pp. 17-45, esp pp. 33-34.
2 See Louis Waller "Regulating Birth Technology" (1998) 7(3) *Res Publica* 18, at 22.
3 *Surrogate Parenthood Act 1988* (Qld); *Family Relationships Act Amendment Act 1988* (SA); *Surrogate Contracts Act 1993* (Tas). The *Substitute Parents Agreement Act 1994*, enacted by the Australian Capital Territory's Legislative Assembly, seeks to facilitate altruistic or voluntary surrogacy agreements: Waller (2 above) at 22.
4 See Waller (1 above) at 33, endnote 124.
5 A comprehensive account and commentary on the work of the IVF Committee is in Waller (1 above). And see Loane Skene *Law and Medical Practice – Rights, Duties, Claims and Defences* (1998) p. 12.
6 See Waller (1 above) endnotes 10-20.
7 Louis Waller "Any Reasonable Creature in Being" (1987) *13 Monash LR* 37, at 53-54; and (2 above) at 20.
8 'Judicial Independence – A Fragile Bastion' (1981) 13 *MULR* 334, at 342.
9 Sec 43(b), and see Article 16(3) of the Universal Declaration of Human Rights, adopted by the General Assembly of the United Nations in 1948.
10 Paras 3.29-3.36.
11 Paras 3.1-3.32.
12 Paras 4.8-4.17.

13 Sections 10(3)(e), 11(3)(e), 11(5)(c), 12(3)(e), 12(5)(c), 13(3)(f), 13(5)(i), 13A(3)(e).
14 Part III, sections 19-23.
15 Section 6(5).
16 Section 6(2)(a).
17 Section 30.
18 Section 29.
19 The *Infertility (Medical Procedures) (Amendment) Act 1987*, introducing section 9A.
20 See Karen Dawson and John Leeton "The regulation of assisted reproductive technology in Australia: issues and solutions" (1995) 163(4) *M J Aust* 204.
21 *Judge and Jurist in the Reign of Victoria* (1959) p. 130.
22 See page 2, above.
23 See, e.g., Belinda Bennett "Posthumous reproduction and the meaning of autonomy" (1999) 23 *MULR* 286, and "Reproductive technology, public policy and single motherhood" (2000) 22 *Syd L R* 625 and see Louis Waller "The cost of treating infertility" (1999) 7(2) *Journal of Law and Medicine*, p. 183.
24 See Waller, note 1 above, at 36.

Chapter 8

Australia – A Federated Structure Of Statutory Regulation Of ART

Helen Szoke[1]

Introduction

The regulation of ART in Australia provides an interesting case study. Australia is a federation, and the Australian Constitution defines the division of powers between the state and federal governments. We are a relatively small country, with a population of only 19 million people, spread across a vast geographical area. Notwithstanding our modest population, we are a country that has been a significant player in terms of the development of assisted reproductive technologies (ART) at a medical and scientific level. The second IVF baby in the world was born in Melbourne, Victoria in 1980 (Johnston et al., 1981). Arising from these early developments in the field of reproductive technology, there is a strong history and tradition of professional self-regulation. The strength of this accreditation program (Reproductive Technology Accreditation Committee – [RTAC]) is further enhanced by the strong financial support which is afforded to infertility treatments in Australia by the federal government, which is evidenced by the provision of a rebate for medically indicated treatments, (and there is now no cap on the number of treatment cycles which are eligible for rebate). There is also a government link with the professional accreditation process, as the Federal Government subsidizes pharmaceuticals for those clinics which hold accreditation by the Fertility Society of Australia's RTAC. The other distinctive feature of the Australian experience is the statutory regulatory experience, with Victoria being the first common law jurisdiction in the world to pass legislation to regulate infertility treatment. Arguably, Australia may be described variously as a rich tapestry of diversity in terms of the regulatory structure, or a patchwork of regulatory stitching lacking cohesion and order.

This chapter provides an overview of the regulatory framework in Australia. In doing so, I will outline three different types of government regulation of reproductive technologies, and look at how these each deal with some of the specific aspects of infertility treatment.

What regulation exists in Australia?

Government intervention will often occur in response to concerns about the consequences or effects arising from progress and development in technology. This is clearly evident with the emergence of the birth technologies in the late 1970s. In Australia, every state government commissioned an enquiry into the use of the technologies, and the Federal Government also canvassed laws and guidelines (Chalmers, 1985; Connor & Kelly, 1984; Demack, 1984; Medical Research Council, 1982; Medical Research Ethics Committee, 1985; National Bioethics Consultative Committee, 1989; National Bioethics Consultative Committee, 1990; NHMRC, 1982; NHMRC, 1983; NHMRC Medical Research Ethics Committee, 1985; NSW Law Reform Commission, 1986; NSW Law Reform Commission, 1987; NSW Law Reform Commission, 1988; WA Parliamentary Committee, 1986; Waller, 1990). The responses to these enquiries differed. Only three state governments moved to legislation to regulate ART. Victoria, South Australia and West Australia have specific legislation regulating assisted reproduction. Queensland, Tasmania, ACT have regulation dealing with surrogacy, and all states have legislation which addresses the legal status of an offspring born as a result of a donor treatment procedure. This diversity may be indicative of the enormous difficulty in governments reaching consensus on the complex moral and value issues posed by the use of reproductive technologies. The split of state and federal powers, as determined by the Australian Constitution, made the development of uniform national legislation difficult.

Table 8.1, in Appendix 1, outlines the current status of legislation in Australia, and how this affects different processes in ART treatment. A summary of ART legislation is shown in Figure 8.1 below.

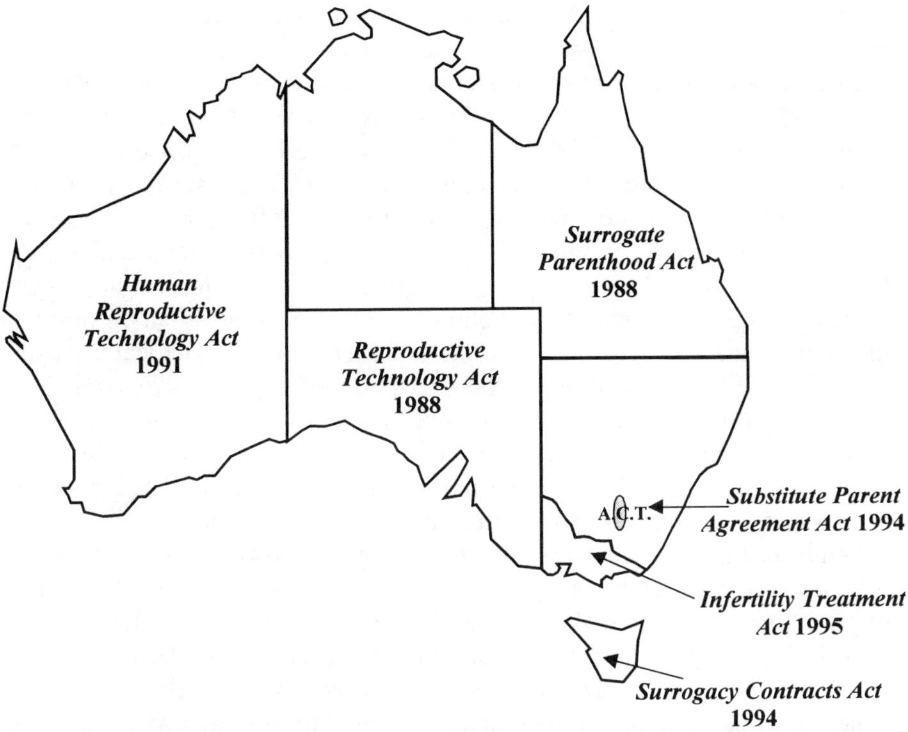

Figure 8.1 Statutory regulation of ART treatment by state – Australia 2001

Continuum of regulatory initiatives

There are different levels of intervention, which can be used to regulate behavior in this area. There are various explanations of the regulatory continuum,[2] but for the purposes of this consideration, I want to look most closely at different types of statutory regulation and their respective benefits and strengths. I will not address the issue of self-regulation by the profession, other than to say that in the Australian experience, in my view, it is important to recognize that peer review and self-regulation, and statutory regulation are not mutually exclusive concepts. The role of self-regulation is to establish clinical and scientific standards, to protect the interests of the patient and to protect the interests of the profession. This is complementary to the process of governments passing laws which address the social, ethical and legal consequences of the use of the technology.

Three states in Australia have passed legislation which generally regulates ART.[3] Statutory responses have also developed in associated areas to deal with the consequences of the use of reproductive technology, such as amendment of existing statutes to ensure that the legal status of children born as a result of donor procedures is clear,[4] or where arrangements associated with IVF surrogacy have been defined by statute.[5] In those states where statutory regulation of ART exists, the regulatory statute operates in conjunction with the profession's self-regulatory structure, the former defining the social and legal considerations and the latter defining the clinical and scientific standards. In those states, where there is no statutory regulation, the guidelines of the Australian pre-eminent medical council in Australia, the National Health and Medical Research Council (NHMRC), addresses the provision of reproductive technology services and the conduct of human embryo research.

For the purposes of this discussion, I have identified three different types of statutory regulation. The first is legislation which identifies bodies to produce guidelines which control institutional or professional behavior through mechanisms such as education programs or funding. This type of regulation is by government guidelines, and is evidenced in the imposition of guidelines, by the NHMRC. The second type is that of legislation which enshrines principles that protect public interest. The principles reflect those priorities that the government has identified as of major community concern. The operational context of the regulation is then delegated to a governing body to determine the application of those principles.[6] This type exists to some extent in the regulatory frameworks that exist in South Australia (SA)[7] and West Australia (WA),[8] where Codes of Practice and Directions determine the basis of operation of clinics, and are able to be reviewed and modified on the recommendation of the governing body.

Finally, there is legislation that enshrines principles, but also defines in the legislation the operational considerations within the context of those principles. This type exists in Victoria, where there is a clear statement of principle, minimal delegation to the governing body and a clear expression of government policy.[9] The administration of this legislation is delegated to a government statutory organization.

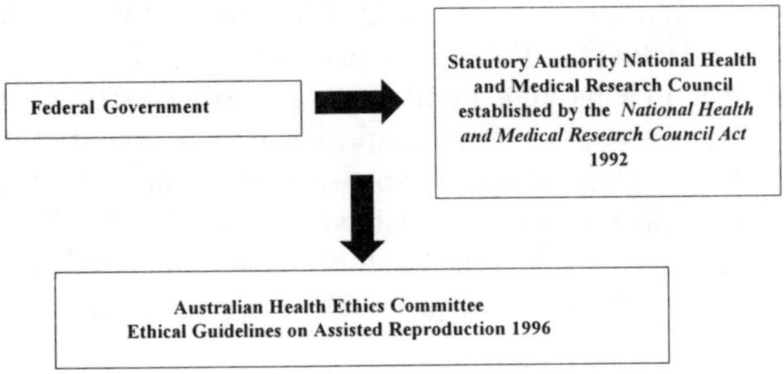

Figure 8.2 Legislation which identifies bodies to produce guidelines

In the area of general regulation of assisted reproduction, the Federal Government has exercised its powers at a number of different levels. In the first instance, it has supported the use of reproductive technology through the subsidization of pharmaceuticals for those clinics accredited through the RTAC accreditation process. Infertility treatment receives a considerable subsidization through the provision of Medicare rebates, through the national health insurance scheme. These rebatable items are for those procedures which the practitioner considers "clinically relevant" (1973).[10] Finally, the NHMRC, Australia's primary medical body, has released guidelines in relation to the provision of infertility services.[11] These guidelines relate to those places which are in receipt of NHMRC grant funding and which have duly constituted institutional ethics committees. The guidelines are deliberately silent on key social issues such as who may receive treatment and surrogacy, and relegate the responsibility for these expressions to state governments, reflecting the understood division of powers. These contentious areas are, to some extent, addressed in the definition of who may receive a rebate for treatment, through the Medicare scheme.[12]

The role of the Federal Government in relation to the development of regulatory frameworks for reproductive technologies has been clearly defined by the Constitutional division of powers between the Federal and State Governments. While the Federal Government does not have uniform regulatory structures through statute, successive Federal governments have provided leadership in certain areas of regulation.[13] More recently, the issue of cloning has galvanized the states' support for the inclusion of a prohibition on reproductive cloning in the Gene Technology Act 2000. The Prime Minister has included the subjects of therapeutic cloning and the experimentation in the area of Embryonic Stem (ES) Cells at the recent Council of Australian Government (COAG) meeting (Wright, 2001). If consensus across state jurisdictions is achieved, it will provide a most significant advance in achieving a uniform regulatory structure in Australia. This is generally viewed as a most positive objective, but not easily attained. To a large extent, it is the expansion of research interests beyond those of infertile people which has enabled even this level of federal/state co-operation.[14] These are issues of high contention and international import. This is most evident in the prohibition of reproductive cloning.

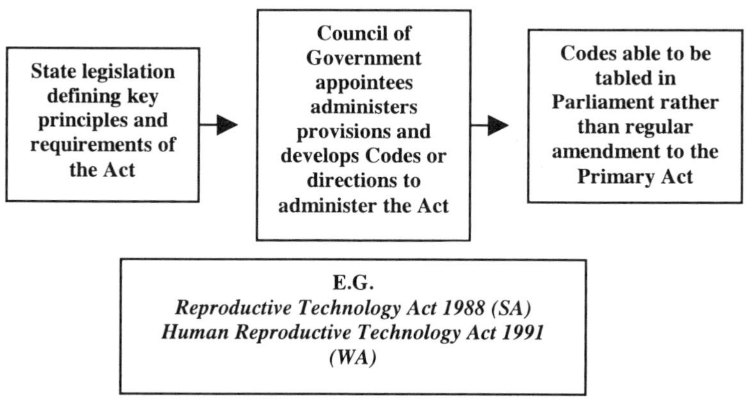

Figure 8.3 Legislation that defines principles but delegates operations to a governing body

Some state governments have opted for the definition of principles enshrined in legislation and the delegation of implementation to codes or directions, which periodically are to be endorsed by the parliament or the parliament's representative through the bureaucratic head of the department. The process requires that the Codes are tabled in Parliament which makes any changes slow and cumbersome, and means that changes can be made to the governing bodies recommendations in Cabinet and Parliament. However the requirement to table the Code is institutionalised, and is still more flexible than an amendment to the legislation. This type of statutory regulation captures the public interest through the articulation of principles, but allows a certain amount of flexibility in relation to its implementation. There are still 'bottom lines' as is evidenced in the SA and WA examples of legislation, where requirements stipulate who may receive treatment, and rules are applied about the use of human embryos and destructive human embryo research. Notwithstanding the clear articulation of principles, it becomes possible to 'check' and monitor the community's response to the imposition of a statutory structure. Legislation, which allows the making of regulations to enhance its operation, or the development of Codes, is more likely to be flexible and responsive than legislation which requires amendment to reflect its intent. In SA in particular, there has been evidence of community consultation around issues associated with donor record keeping, statutory requirements in terms of determining who is fit to parent, the application of preimplantation genetic diagnosis (PGD) procedures (SACRT, 2000). Review and consultation in this area is never easy for governments however, as is evidenced by the recent review of state legislation undertaken by the WA Parliament, which has yet to reach final resolution, (WA Legislative Assembly, 1999) and the long running review in NSW, where since 1997, consideration has been given to the *Human Tissue Act 1995*, including proposals for regulation of assisted reproduction (NSW Health, 1997). This is evidence of the complexity of regulation in areas of high controversy, and concern that the interests of not only those receiving treatment, but of the person born, and the community overall, continue to be of significance.

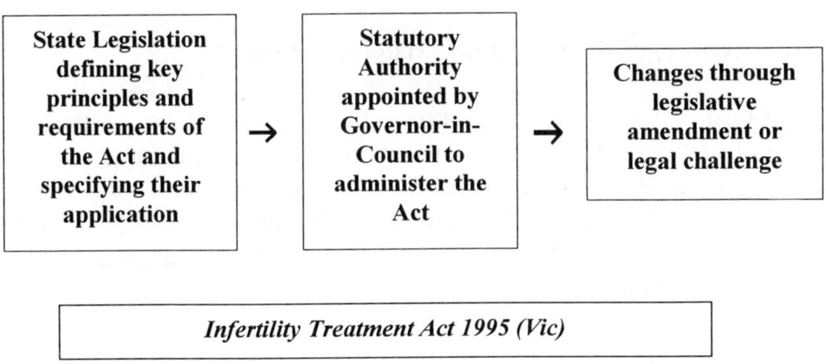

Figure 8.4 Legislation that defines principles and also enshrines operational aspects

The final type of statutory regulation is that where both the principles and the 'rules of conduct' are expressed in legislation, which is comprehensive and then delegated to a governing body to administer compliance. The

Australian version of this exists in Victoria, with the passage of the Infertility Treatment Act 1995, and the creation of the Infertility Treatment Authority. Principles are clearly defined in descending order of importance, with the interests of the person born of paramount concern, and the interests of the infertile couple incorporated but of a lesser status.[15] The Victorian Act clearly enshrines a regulatory framework which expresses a community view about who should receive treatment, how children born of the procedures should be treated and what the status of the human embryo is. This legislation had bipartisan support until 2001, as it was passed without major amendment in 1995. Bi-partisan support for the Act ceased late in 2001, when the Opposition introduced an amendment into the upper house of the State Parliament proposing a new definition of who may access infertility treatment.[16]

The Act has been amended to include *de facto* couples for treatment.[17] The Act was further amended to allow the use of embryos posthumously and to establish a voluntary birth register for pre-1988 donor procedures.[18]

Challenges have been made as a result of legal action, resulting in the order of the Federal Court which removed the marriage requirement from the state legislation.[19] Challenges have also come from the scientific and clinical community who are concerned about the impact of the legislation on their research initiatives and treatment options.

How do these types of regulatory structures impact?

The impact of these different types of regulatory structures can be assessed by looking at three different aspects of infertility treatment and research. The question of who may be treated has been a matter of legal contest and policy reform in Australia. This is explored below, along with the statutory protections for donor procedures and finally the regulation of human embryo experimentation.

Who may be treated?

In those cases where legislation identifies bodies to produce guidelines as is evidenced in the NHMRC, the guidelines are deliberately silent about who may access infertility treatments, leaving this to the individual state jurisdictions, where they exist, to address. In practice this means that where state legislation exists, that will determine the access requirements. Where such legislation does not exist, then the decision about who may be treated will be left to the individual clinic, or in some cases may be addressed by the institutional ethics committees associated with institutions where clinics are based. The financial arrangements at a federal level may also have some influence on who may be treated, in that only clinically indicated treatments are subject to the national health scheme, Medicare, rebate.

In those states where specific legislation regulates infertility treatment, the definition of who may be treated is incorporated into the core legislation. In South Australia, the legislation specified heterosexual couples as the eligible treatment group. This was challenged in a human rights hearing[20] and resulted in the requirement for evidence of infertility to enable treatment to be provided. In West Australia, the requirements for accessing IVF are stipulated in the legislation and limited to married or *de facto* couples, but on the matter of who may access DI treatment, the legislation is silent. Therefore, there is no barrier to single women accessing such treatment. The WA Act is currently under review, and also influenced by the West Australian Government Gay and Lesbian Reform Package, which proposes to remove all gender specific references in the Artificial Conception Act 1985. This means that where a woman is in a *de facto* relationship with another woman, and undergoes an artificial fertilization procedure, with the consent of her partner, then the de facto partner is the legal parent of the resulting child.[21]

In Victoria, where legislation is more prescribed, there has been a most public discussion of who may receive treatment. Since 1995, Victorian legislation has gone through a number of manifestations – first defining that treatment would only be available to married couples, then an amendment to the Act in 1997, before proclamation of the Act, and in response to another human rights hearing, the Act allowed access to heterosexual *de facto* couples.[22] In 2000, the now famous case of McBain v State of Victoria[23] resulted in the ruling that the 'marriage requirement' could no longer be imposed as a barrier to treatment. The response following this case from the Infertility Treatment Authority was to reinforce that the Act was passed by the Parliament to regulate infertility treatment in the State of Victoria, and therefore evidence that the woman "... is unlikely to become pregnant except with the use of a treatment procedure ...",[24] and thus that clinical infertility remained a requirement.

The aftermath of the case is worth quickly describing as it is evidence of the importance of the question of who may receive infertility treatment in the Australian public policy context. The political response was immediate. The announcement of the Court coincided with the National Conference of the Australian Labor Party, in opposition at Federal level.

The Prime Minister intervened and proposed an amendment to the Federal Sex Discrimination Act. This move, if passed by the Federal Parliament, would allow States an exemption from the provision of the Sex Discrimination Act 1984 in relation to who may access IVF services, allowing states to prescribe these services for heterosexual couples, if they so chose.

This was the converse of the normal tradition of a federal structure, where the national legislation constitutionally has precedence over state legislation. It was also alleged that the Government's actions challenged Australia's involvement in international treaties and agreements related to the protection of women's rights (Lees, 2001). Commentators at the time reflected that this was a move by an ailing Federal Government to divert attention from the Opposition and their major policy making forum (Hudson, 2000). Notwithstanding, the announcement resulted in the introduction of a Bill to amend the Sex Discrimination Act 1984 to allow the State legislatures to impose such restrictions to treatment.

A subsequent Senate Select Committee recommended that the Act should *not* be amended.[25] The Bill remains at second reading stage in the Senate during 2002.

In November 2001, guidelines proposed by the Victorian Infertility Treatment Authority addressing the eligibility status of those women who were unable to have heterosexual intercourse, became the subject of enormous media and political reaction, with public spokesperson claiming that reproductive technology should only be for medically indicated infertility treatments. The Opposition government in Victoria introduced an amendment to the *Infertility Treatment Act 1995*. The intent of this amendment is to ensure that only medically infertile women may be eligible for treatment. This amendment may be the subject of consideration in the Parliament, later in 2002.

Donor procedures

The NHMRC guidelines specify that "children born from the use of ART procedures are entitled to knowledge of their biological parents. Any person, and his or her spouse or partner, donating gametes and consenting to their use in an ART procedure where the intention is that a child may be born must, in addition to the information specified in this section, be informed that children may receive identifying information about them" (NHMRC, 1996). How this is enforced and monitored is not clear. Certainly anonymous donation continues to be offered in NSW, Queensland, ACT and Tasmania, and there is no mechanism currently in place which provides statutory protection to birth origin information, other than state statutory requirements in relation to retention of medical records. There remain concerns in those states about the impact of disclosure on donor recruitment, and a lack of commitment to the importance of ensuring the availability of birth origin information, even though these states have such provisions enshrined in legislation in the adoption area. The decision about accessing birth origin information is therefore left to specific requests made to institutional ethics committees.

WA and SA are both states in transition. WA, in the review of its legislation, and SA through its consultative processes, have adopted a policy position supporting the protection of birth origin information and its access by the parties involved with donor procedures. The SA process has been particularly transparent and consultative, providing the opportunity for education and raising community awareness about the issues associated with donor procedures, while carefully moving towards a system of regulatory protection and control of such information.[26] A proposal to amend the SA legislation to allow the establishment of a Register is currently being considered. This matter has been reviewed in WA as part of the overall review of legislation regulating ART in that state.

In Victoria, the statutory protections in relation to accessing birth origin information are enshrined in the primary legislation. This includes the process of counseling, information giving, consent, and then the record keeping requirements. Victoria has required that all donor procedures be reported to a Central Register since 1988, and since 1998, offspring have been prospectively provided with the right to access identifying information about their donor, on turning the age of 18 years. All donors have consented to the use of their gametes on this understanding. The Registers are maintained by the Authority, and the management of any release of information is carefully regulated by that body.

The matter of access to birth origin information is a contentious issue. There are very few examples internationally where the concept of protection of access to birth origin information is embraced by the legislature. The Australian experience highlights that in this area the Victorian legislation in fact offers a strong statutory protection to people involved with donor procedures. It should also be noted that the 'trend' is to extend these protections in other state jurisdictions. NSW, in its review of legislative requirements, has this as a consideration, and as stated above WA and SA have incorporated a policy position supporting such protections. What happens in Victoria, and ultimately in other states, will provide an important model of the statutory protections which can be offered in this area internationally.

Embryo experimentation

The provisions in relation to embryo experimentation have been fundamentally challenged by the advances in the science and the technology. These challenges have been posed predominantly by the reported developments in relation to embryonic stem cell (ES) research, and the benefits which are claimed to be accrued to the broader population, beyond the interests of the infertile. The current status of the regulation of human embryo research in Australia is outlined below.

The NHMRC has stated thus – "Research involving early human embryos raises profound moral and ethical concerns. There are differences of opinion amongst Australians regarding the moral status of the human embryo, particularly in its early stages of development. At the present time these differences cannot be resolved" (NHMRC, 1996). Destructive human embryo research is allowed in exceptional circumstances, where significant advances in knowledge are anticipated, a limited number of embryos are used and the people who provided the gametes have given their consent (NHMRC, 1996). However, as outlined above, the issue of embryo research has been placed before the Federal Parliament, in the *Human Cloning and Embryo Experimentation Bill 2002*, proposing national regulation.[27]

The challenge to achieving national regulation in this area lies in what I would call the bottom line provisions in those states which have legislation. In Victoria, SA and WA, destructive research utilizing a human embryo is prohibited. In all of those states, the definitions within the legislation did not contemplate ES cells and research, and the importation of ES cell lines into SA and Victoria has not fallen within the current statutory reach.

The success or otherwise of achieving a national regulatory scheme rests on consensus between the states which have restrictive provisions relating to human embryo research, and those which are unregulated except by NHMRC Guidelines. This may be a difficult consensus for the Federal Government to achieve.

Conclusion

From a regulators perspective, I believe there is a complementary relationship between professional self regulation and statutory regulation – those matters of risk management in technical, clinical and scientific matters are properly the province of the profession. I concede the definition of the boundaries is a little tricky. The Australian experience is important as it maintains the responsibility for clinical and professional standards with those people responsible for providing the service. The expression of public policy thus resides with government.

The current developments in the research areas utilizing human embryos have, in my view, provided a potential galvanizing influence for some uniformity to be achieved in relation to national regulation in Australia in this area. Such uniformity will be a challenge however, as at least three states have strong legislative protections for human embryos and their use beyond the use in a treatment procedure. As the medical and scientific benefits of research using human embryos expand beyond solely the interests of the infertile, there is a greater community of interest in both the regulation and the outcome of such a resource.

Statutory regulation offers strong protections in those areas where the direct interests of offspring are involved, as is evidenced in the donor procedure record keeping requirements in Victoria and soon to be introduced in South Australia. This area I believe provides an important model for other jurisdictions. Victoria is already building up a working experience of donor linking and an understanding of the issues which arise. This will be important information internationally to inform how donor treatments are provided.

The question of who may access treatment is problematic. Different legislative regimes have been challenged where they are restrictive, and then challenged again when they are liberalised. The Australian Federal Government has a "card up its sleeve" with the proposed amendment to the Sex Discrimination Act which would allow states to make their own laws. The matter awaits possible legislative amendment in Victoria. The forthcoming federal review of regulation may offer some resolution to this issue.

In Australia, the current regulatory models and their diversity may in fact meet the description of a rich tapestry in an area of high contention and division. These are different types of statutory regulation in the Australian experience. They offer choice and diversity for Australia's population. This diversity may also frustrate and confuse patients. The Australian experience is one of a federated governmental structure, exploring the need and the possibility of achieving a uniform structure for this jurisdiction. Sir Winston Churchill once said that "headmasters have powers at their disposal with which Prime Ministers have never yet been invested". This difficulty clearly applies in the Australian experience, where a uniform regulatory structure cannot be applied, but must be negotiated.

Notes

1 Those aspects of this manuscript, which are opinion or an evaluation, are not the opinion or the evaluation, express or implied, of the Victorian Infertility Treatment Authority.
2 See for Example Blank's continuum of regulation (Blank, 1984).
3 *Reproductive Technology Act* 1988 (SA).
Human Reproductive Technology Act 1991 (WA).
Infertility Treatment Act 1995 (VIC).
4 *Status of Children Act* 1974 (TAS).
Artificial Conception Act 1984 (NSW).
Family Relationships Act 1984 (SA).
Family Relationships Amendment Act 1984 (ACT).
Status of Children Act 1984 (VIC).
Artificial Conception Act 1985 (WA).
Artificial Conception Act 1985 (QLD).
5 *Surrogate Parenthood Act* 1988 (QLD).
Surrogacy Contract Act 1993 (TAS).
Substitute Parent Agreement Act 1994 (ACT).
These Acts were passed in jurisdictions where no other broader regulatory framework existed for reproductive technologies. In the case of the ACT, the Act facilitates the provision of altruistic gestational surrogacy services, and through the administrating of the statute, has now become the first Australian government jurisdiction to also address the legal issues associated with the child born as a result of an altruistic surrogacy arrangement. Division 3.2 of the *Artificial Conception Act* 1985 states how a parentage order can be obtained in relation to certain children born as a result of surrogacy agreements.
6 The best example of this exists in the United Kingdom, through the enactment of the *Human Fertilisation and Embryology Act* 1990. This Act requires the development of Codes of Practice, but also has greater capacity to allow the making of regulations to ensure the law keeps abreast of technological developments and community acceptance and understanding.
7 *Reproductive Technology Act* 1988 (SA).
8 *Human Reproductive Technology Act* 1991 (WA).
9 *Infertility Treatment Act* 1995.
10 *Health Insurance Act* 1973 (Cth).
11 NHMRC (1996), *Ethical guidelines on assisted reproductive technology.*
12 Women who seek reproductive technology treatment for non-medical reasons have to pay for it themselves.
13 Such as the definition of the status of the child from donor procedures, or the clarification of surrogacy in the Australian context, follows the demise of the National Bioethics Consultative Committee, and the agreement of the Australian Health and Welfare Ministers meeting to oppose facilitation of altruistic surrogacy. National Bioethics Consultative Committee, *Surrogacy. Reports 1 and 2* (NBCC ACT, 1990); Family Law Council, *Creating children: A uniform approach to the law and practice of reproductive technology in Australia* (Family Law Council, Canberra ACT, 1985).
14 See for example the Roy Morgan International Poll which shows strong support for couples donating excess embryos, with consent, to research rather than discarding them, and that the greater proportion of respondents approved of therapeutic cloning or nuclear transfer. Roy Morgan International, "Embryonic Intent" (2001) The Bulletin 34.

15 Section 5 of the *Infertility Treatment Act* 1995 states that "It is the Parliament's intention that the following principles be given effect in administering this Act, carrying out functions under this Act, and in the carrying out of activities regulated by this Act – (a) the welfare and interests of any person born or to be born as a result of a treatment procedure are paramount; (b) human life should be preserved and protected; (c) the interests of the family should be considered; (d) infertile couples should be assisted in fulfilling their desire to have children". The Act requires that the principles must be applied in the descending order of importance as written in the Act.

16 Infertility Treatment (Further Amendment) Bill 2001. This Bill was introduced in response to concerns about proposed guidelines for 'psychological infertility' which were released for consultation by the Infertility Treatment Authority. See ITA News, December 2001. The effect of the proposed amendment is to clearly define the eligibility for treatment as some form of clinical infertility in the strictest sense, and to prevent lesbian women who are not clinically infertile from receiving treatment.

17 This occurred in 1997, before full proclamation of the Act and in response to legal proceedings brought against a practitioner and hospital.

18 *Infertility Treatment (Amendment) Act* 2001.

19 McBain v State of Victoria & Ors, [2000] FCA 1009.

20 M W and Ors v Royal Women's Hospital & Ors, Human Rights and Equal Opportunity Commission, H96/26, 96/33 and 96/48, A. Kohl, Melbourne, 12 March 1997.

21 The Acts Amendment (Lesbian & Gay Law Reform) Bill 2001 (WA).

22 M W and Ors v Royal Women's Hospital & Ors, Human Rights and Equal Opportunity Commission, H96/26, 96/33 and 96/48, A. Kohl, Melbourne, 12 March 1997.

23 McBain v State of Victoria & Ors [2000] FCA 1009.

24 *Infertility Treatment Act* 1995. Section (3)(a).

25 Senate Legal & Constitutional Legislation Committee. "Inquiry into The Sex Discrimination Amendment Bill (No. 1) 2000". February 2001. Commonwealth Government, Canberra.

26 See for example (SACRT, 2000).

27 House of Representatives Standing Committee on Legal and Constitutional Affairs, 2000. Inquiry into The Scientific, Ethical & Regulatory Consideration Relevant to the Cloning of Human Beings (Canberra, Parliament House).

References

Acts of Parliament

Health Insurance Act 1973 (Cth).
Status of Children Act 1974 (TAS).
Artificial Conception Act 1984 (NSW).
Family Relationships Act 1984 (SA).
Family Relationships Amendment Act 1984 (ACT).
Status of Children Act 1984 (VIC).
Artificial Conception Act 1985 (QLD).
Artificial Conception Act 1985 (WA).
Reproductive Technology Act 1988 (SA).
Surrogate Parenthood Act 1988 (QLD).
Human Reproductive Technology Act 1991 (WA).

Surrogacy Contract Act 1993 (TAS).
Substitute Parent Agreement Act 1994 (ACT).
Sex Discrimination Amendment Bill (No 1) 2000 in: Commonwealth (Ed) *A Bill for an Act to amend the Sex Discrimination Act* 1984.
Infertility Treatment (Amendment) Act 2001, Infertility Treatment (Further Amendment) Bill 2001 (VIC).

Government Reports

Chalmers, D. (1985) Committee to Investigate Artificial Conception and Related Matters (Hobart, Tasmania).

Connor, A.F. & Kelly, P. (1984) Report of the working party on in vitro fertilisation and artificial insemination by donor (South Australia).

Demack, J. (1984) Report of the Special Committee appointed by the Queensland Government to enquire into the laws relating to artificial insemination, in vitro fertilisation and other related matters (QLD).

House of Representatives Standing Committee on Legal and Constitutional Affairs (2000) Inquiry into the Scientific, Ethical and Regulatory Considerations Relevant to Cloning of Human Beings (Canberra, Parliament House).

Medical Research Council (1982) Research related to human fertilisation and embryo (London, Medical Research Council).

Medical Research Ethics Committee (1985) Embryo Donation by Uterine Flushing. Interim Report on Ethical Considerations (Canberra, ACT, NHMRC).

National Bioethics Consultative Committee (1989) Reproductive Technology. Record Keeping and Access to Information. Birth Certificates and Birth Records of Offspring Born as a Result of Gamete Donation (Canberra, ACT, NBCC).

National Bioethics Consultative Committee (1990) Surrogacy. Reports 1 and 2 (Canberra, ACT, NBCC).

NHMRC (1982) NHMRC Statement on Human Embryo Experimentation (Canberra, ACT, NHMRC).

NHMRC (1983) Ethics in medical research involving the human foetus and human foetal tissue (Canberra, ACT, NHMRC).

NHMRC (1996) Ethical guidelines on assisted reproductive technology (Canberra, ACT, NHMRC).

NHMRC Medical Research Ethics Committee (1985) Embryo Donation by Uterine Flushing. Interim Report on Ethical Considerations (Canberra, ACT, NHMRC).

NSW Health (1997) Review of the Human Tissue Act 1983. Discussion Paper. Assisted Reproductive Technologies (Sydney, NSW, NSW Health).

NSW Law Reform Commission (1986) Human artificial insemination (Sydney, NSW, NSW Law Reform Commission).

NSW Law Reform Commission (1987) In vitro fertilisation (Sydney, NSW, NSW Law Reform Commission).

NSW Law Reform Commission (1988) Surrogate Motherhood (Sydney, NSW, NSW Law Reform Commission).

SACRT (2000) Donor Conception in *SA Council on Reproductive Technology Quarterly Bulletin.*

Senate Legal and Constitutional Committee. "Inquiry into The Sex Discrimination Amendment Bill (No. 1) 2000". February 2001. Commonwealth Government, Canberra.

WA Legislative Assembly (1999) Select Committee on the Human Reproductive Technology Act 1991 (Perth, WA Parliament).

WA Parliamentary Committee (1986) Report of the Committee appointed by the Western Australian Government to Enquire into the Social, Legal and Ethical Issues Relating to In vitro Fertilisation and its Supervision (Perth, WA).

Waller, L. (1990) Consolidated Reports of the Victorian Inquiry into IVF and Related Issues. 2nd Reprint (Melbourne, Vic, Committee to Consider the Social, Ethical and Legal issues arising from in-vitro fertilisation).

Books & Articles

Blank, R.H. (1984) *Redefining Human Life: Reproductive Technologies and Social Policy* (Boulder, Colorado, Westview Press).

Johnston, I., Lopata, A., Speirs, A., Hoult, I., Kellow, G. & Du Plessis, Y. (1981) In vitro Fertilisation: the Challenge of the Eighties, *Fertility and Sterility*, Vol. 36, No. 6, December 1981.

Szoke, H. (1999), "The Regulation of Assisted Reproductive Technologies; The State of Play in Australia." In *Controversies in Health Law*. Freckelton, I. and Petersen, K. The Federation Press: Annandale.

Media

Hudson, P. (2000) National split on IVF for singles, lesbians *The Age* (Melbourne).

Lees, M. (2001) IVF breeds dissent *Courier Mail* (Brisbane).

Roy Morgan International (2001) Embryonic Intent, *The Bulletin* (July 24), pp. 34.

Wright, L. (2001) No decision on controversial stem-cell research. Human cloning outlawed. *Canberra Times* (Canberra).

Legal Cases

M W and Ors v Royal Women's Hospital & Ors, Human Rights and Equal Opportunity Commission, H96/26, 96/33 and 96/48, A. Kohl, Melbourne, 12 March 1997.

McBain v State of Victoria & Ors [2000] FCA 1009.

Table 8.1 Guidelines and legislation – Assisted Reproductive Technology – Australia (November 2001)

Legislation	NHMRC	NSW	Victoria	Queensland	South Australia	West Australia	ACT	Northern Territory	Tasmania	RTAC
Specific Legislation	National Guidelines produced for Institutional Ethics Committees [in process of revision Nov 2001]	NIL *Artificial Conception Act* 1984 "Review of Human Tissue Act Discussion paper. 1997" - Dept Health	*Infertility Treatment Act* 1995 *Infertility Treatment (Amendment) Act* 2001 *Infertility (Medical Procedures) Act* 1984	NIL *Surrogate Parenthood Act* 1988	*Reproductive Technology Act* 1988 Exemption from State Sex Discrimination Act.	*Human Reproductive Technology Act* 1991 This Act is silent on Donor Insemination. Parliamentary Review in progress.	NIL *Substitute Parent Agreement Act 1994*	Guided by South Australian *Reproductive Technology Act 1990*	*Surrogacy Contracts Act* 1994	RTAC Guidelines Produced by Fertility Society of Australia
Access to Services	No specific Guidelines	IEC Decision. Access by *de facto* couples, married couples, single women and same sex couples.	Federal Court Case 2000 removed 'marriage requirement' Must be clinically infertile.	Decision made by individual practitioner.	De facto and legally married and single women and same sex couples. Infertility a requirement.	Heterosexual *de facto* and legally married couples. DI exempt from Act.	IEC decision	SA legislation	IEC decision	N/A
Selection Criteria for Access to Services	None specified	None specified	Evidence that a woman is unlikely to become pregnant other than by a treatment procedure; or to prevent the transfer of a genetic abnormality or disease.	None specified	Infertility prevention of a genetic defect being transferred to child; and requirements of parenting code to be met.	Infertile and married or *de facto* for IVF.	None specified	SA Legislation	None specified	N/A
Storage of Embryos	Ten year limit recommended	NHMRC guidelines	Gametes 10 year limit, embryos 5 year limit. Can apply for extension of storage period.	NHMRC Guidelines	Ten year limit for embryos. No provision for extension. Required to contact couples every 12 months.	Consents to be renewed every 3 years for embryos and five years for gametes. Can apply for extensions. Maximum of 15 years storage overall.	NHMRC Guidelines	SA	NHMRC Guidelines	10 year storage limit
Record Keeping for Donor Offspring	Donor offspring have a right to access records about their biological past.	Status of paternity specified presuming the donor not to be the father for all	Central Register established to record all information about donors and	No specific provisions - no records required to be made accessible. IEC	Licensee to maintain records of donor. Access to non identifying	Register to be kept by the Commissioner of Health. Records to be kept by	IEC	SA Legislation	IEC	Recommends the information be made available

Legislation	NHMRC	NSW	Victoria	Queensland	South Australia	West Australia	ACT	Northern Territory	Tasmania	RTAC
		purposes.	offspring. Offspring have a right to access information. Other parties may access with consent. Voluntary Register to allow information exchange.	decision.	information is available for offspring. Currently proposing amendment to Act to set up Birth Origin registers.	licensees. Record the identity of offspring and donor, and identity of parents of the child. Confidentiality of information assured.				
Legislative Definition of Parentage in Donor Procedures	Federal *Family Relationships Act* 1975 (Amendment Act 1984) Woman who gives birth is the mother. Man who donates sperm is not presumed to have caused the pregnancy.	*Artificial Conception Act* 1984 Ibid.	*Status of Children Amendment Act* 1984 Ibid.	*Artificial Conception Act* 1985 Ibid.	*Family Relationships Amendment Act* 1984 Ibid.	*Artificial Conception Act* 1985	*Artificial Conception Act* 1986	*Status of Children Amendment Act* 1995	*Status of Children Act - Amendment* 1985	N/A
Informed Decision Making	Specific provisions outlined in guidelines e.g. consent in writing, medical practitioners responsibility, adequate information, providing for counseling.	NHMRC Guidelines	Specific requirements for consent outlined in Conditions for Licensing and in the Act. Prescribed matters to be covered in counseling specified in regulations. Counseling mandatory.	NHMRC Guidelines	Specific provisions in relation to consent and consent forms. Information Statement Standards specified - written information which must be distributed.	Specific requirements outlined in the Act. Guidelines in relation to information, consent and counseling. Differentiate information giving, counseling and consent.	NHMRC guidelines.	SA	NHMRC guidelines	
Research	Guidelines specifying nature of embryo research - therapeutic research subject to IEC approval. Destructive research can be approved in	NHMRC guidelines	Ban on destructive research specified in the Act. Act does not cover ES cells once formed.	NHMRC Guidelines	Act specifies no destructive research on embryos. Non-destructive research requires approval of Reproductive	Therapeutic or specific approved research on embryos. Prohibition on keeping embryo longer than 14 days.	NHMRC guidelines	NHMRC	NHMRC guidelines	

Legislation	NHMRC	NSW	Victoria	Queensland	South Australia	West Australia	ACT	Northern Territory	Tasmania	RTAC
	exceptional circumstances.				Technology Council. Act does not cover ES cells once formed.	Prohibition on destructive research. Act does not cover ES cells once formed.				
Surrogacy	Not included in guidelines.	No explicit laws BUT: Birth parents registered as parents. Adoption - do not facilitate adoption in surrgoacy - refer the matter to the Family Court.	Illegal to advertise or receive payment for surrogacy arrangement. Birth parents registered as parents. Must meet eligibility requirements of Act to be treated. Agreements void.	*Surrogate Parenthood Act* 1988. Offence to make or seek surrogacy arrangement. Surrogacy Agreements unenforceable.	Contracts non legal and void, but no penalties imposed. Rely on Family Relationships Act 1975 to define parentage of children which would pose an impediment.	No specific legislation. Rely on common law provisions. Also view that surrogacy is against public policy view and therefore any contracts are void.	*Substitute Parents Act* 1994 allows altruistic surrogacy. Act prohibits commercial arrangement or facilitating a commercial arrangement.	SA	*Surrogacy Contracts Act* 1993. Offence to make or seek surrogacy arrangement. Surrogacy Agreements unenforceable. Offence to offer support or assistance for such an arrangement.	

Chapter 9

The New Zealand Way: ART Within An Ethical Framework

Rosemary De Luca

Introduction

The field of assisted human reproduction is largely unregulated in New Zealand. This situation, however, is expected to change in the near future. Early in 2002 when this account of New Zealand's approach to regulating assisted reproductive technologies (ART) is being prepared for publication, legislation is at the Select Committee stage. Themes in the proposed legislation relate to benefits and harms, and also information, access and identity for those who become involved in applications of the technologies. The proposed legislation identifies prohibited practices and mechanisms for licensing and ethical review. It is expected to balance the rights of infertile women and men; the rights of children born as a result of ART; the exigencies of rapid technological advances; the caution of the more morally conservative among New Zealanders; the concerns of some community sectors about challenges to the norms of nature; and the widely held expectation that society will take advantage of what science can offer to improve the well-being of its members. One could be justified in describing this as a seemingly impossible task.

The account which follows describes how, in the absence of legislation specifically designed for the purpose, constraints have been formally exercised in the provision of ART services in New Zealand in the last decade. During this time, fertility specialists have kept up with international developments and offered an increasingly wide range of services. Since 1993, ethical review of innovative treatment and research provided by the National Ethics Committee on Assisted Human Reproduction (NECAHR) (formerly the Interim National Ethics Committee on Assisted Reproductive Technologies), the guidance offered by professional codes, the requirements of Australia's Reproductive Technologies Accreditation Committee (RTAC) on which New Zealand is represented, the responsible practice of fertility specialists, and a scattering of legislation written for other purposes have somehow resulted in cautious and sensible progress in the field of assisted human reproduction. Issues dealt with by the Ethics Committee include the storage, use, disposal and retrieval of human gametes, research involving embryos, donation of stored embryos, pre-implantation genetic diagnosis, and surrogacy.

Most of this chapter focuses on one area of service provision using reproductive technologies: in vitro fertilization to allow a surrogacy arrangement to proceed. It tracks the progression from the Ethics Committee's refusal to give ethical approval for a proposed surrogacy arrangement in 1993 to a situation, almost ten years later, which involves a complex but workable set of procedures whereby over fifteen approvals have been given. Regard for the rights of all parties, the interests of potential children, respect for the autonomy and privacy of individuals, active acknowledgement of Treaty of Waitangi obligations, regard for the diverse views in New Zealand's multi-cultural society, and respect for the professionalism of service providers underlie the ethical review procedures.

The approach to surrogacy which is described in what follows exemplifies a framework of control which rests, for the most part, on ethical review. This occurs in the absence of specifically designed legislation and in the context of existing legislation interpreted to fit new situations which would hardly have been imagined when the laws were drafted. The various parties, through discussion, negotiation and information sharing, work to reach consensus about how an application of the technologies might proceed in an ethical manner. Towards the end of the chapter there is a brief summary of attempts to develop policy and legislation in the ART field, and of the status quo early in 2002.

A committee on a slippery slope or a committee with a legitimate ethical theory?

In 1993 and again in 1994, the Interim National Ethics Committee on Assisted Reproductive Technologies declined to give approval to an application for ethical review from a fertility clinic that proposed to offer non-commercial surrogacy using in vitro fertilization (IVF) as a treatment for infertility. It was the Committee's view that the woman who was to have the fertilized ovum implanted in her uterus, gestate and give birth would find it so hard to foresee 'the emotional trauma associated with giving up the child' that she would be very unlikely to be able to give informed consent, or to consent with an ethically acceptable level of knowledge (*Non-Commercial Surrogacy by Means of In Vitro Fertilization*, 1995, p. 10).

Further, the risks inherent in the proposed surrogacy arrangement influenced the Committee's decision. The Committee wrote in its report:

> There are significant medical and emotional risks associated with IVF surrogacy, including the use of drugs to induce ovulation, and the low success rate. There are also the physical risks associated with pregnancy itself. The risks to the child include the disruption of bonding which may occur during gestation, genealogical bewilderment associated with adoption and the possibility of a custody conflict if the parties do not agree after the birth. There is also the possibility that, for example, in the case of the child with a disability, neither the birth mother nor the infertile couple will wish to keep the child. Even if no money changes hands, surrogacy can lead to the commodification of children, thus breaching an important social value regarding the inherent rather than the instrumental worth of persons (pp. 9-11).

The Committee did not close the door on surrogacy, however, and agreed to re-visit the matter at a later time.

Subsequently, in response to a further application for ethical review in 1997, NECAHR gave approval to surrogacy in principle but clearly specified that it was to be altruistic and non-commercial, and to occur only in situations where infertility was caused by a medical condition. Various criteria were to be met in relation to informed consent and the well-being of all parties involved, including the potential child, both at the time of the surrogacy arrangement and in the future. The Committee also stipulated that application for ethical review was to be made for all subsequent surrogacy proposals on a case-by-case basis. In the following four years the Committee approved over fifteen surrogacy arrangements, all of which met the requirements it had stipulated.

In the four years, guidelines were developed for fertility clinics to follow in making application to the Committee. These were at first referred to as 'evolving criteria'. Later, they assumed the status of draft guidelines, becoming the *Draft Guidelines for Non-commercial Altruistic Surrogacy using IVF as Treatment*.

They were based on issues that arose and were addressed as applications were considered. Fertility clinics offered comment from time to time. New Zealand's Human Rights Commissioner and the Commissioner for Children both reviewed the draft guidelines. In 2001, the draft guidelines were circulated for public consultation. With a small number of minor changes, they will continue as the officially sanctioned operational document for providers of fertility services. They will, however, retain their draft status until legislation clarifies the position with regard to the regulation of assisted reproductive technologies.

An obvious question is what made the Committee change its mind. Some people suggest that the composition of the Committee had changed and that the members in 1997 held more liberal views than those in 1993 and 1994. Some claim that the practice of surrogacy was generally becoming more familiar; others that the members of the Committee had become desensitized to assisted reproduction. Still others attribute the Committee's change of stance to pressure brought to bear by an appeal committee which had been asked for a second opinion and was critical of the 1993 and 1994 decisions. There is probably some truth in all of these explanations.

Perhaps the single most influential factor had to do with process. Not all the possible permutations of a surrogacy arrangement which met the three provisions stipulated (altruistic and non-commercial, and to occur only in situations where infertility was caused by a medical condition) could be worked out from the beginning. It became clear to the Committee that a case-by-case approach with evolving criteria on which to base decision-making offered a way forward.

Furthermore, the Committee adopted an approach which involved professionals reporting to the Committee. Fertility and other specialists, lawyers and counselors would work with their clients, paying due regard to the criteria to be addressed in an application. In reports to the Committee they would describe discussions with clients in general terms, and give assurances that issues had been fully explored and requirements met. This meant that members of the Committee did not have to become unduly involved in the details of people's private lives.

A case-based or casuistic approach to ethical review is not new, although for quite some time it was out of favor with ethics theorists. However, along with others, Beauchamp and Childress (1994), authors of the well-known medical ethics textbook, *Principles of Biomedical Ethics*, now acknowledge casuistry as a respectable ethical theory.

Jonsen and Toulmin (1988) describe the revival of casuistry, the definition of which includes the application of 'general rules to particular cases with discernment'. In relation to difficulties encountered in establishing the general rules, they cite the experience of the National Commission for the Protection of Human Subjects of Biomedical and Behavioral Research in the United States. At first, this Commission tried to develop general statements of ethical principle which would guide biomedical and behavioral research. They found this difficult, and eventually developed 'moral taxonomies for classifying the morally significant similarities and differences between the various kinds of research' (p. 6). They formulated the classification criteria for the taxonomies in a very practical way from existing research proposals, having been unable to reach agreement on general principles without examples to ground them.

The National Ethics Committee on Assisted Human Reproduction took a similar approach. It focused on cases as a way to articulate principles, rather than trying to produce guidelines without any reference points. The three provisions (altruistic and non-commercial, and to occur only in situations where infertility was caused by a medical condition) characterize the casuists' paradigm case and provide the basic 'rules'. Discernment is exercised with reference to the guidelines.

Ethical themes within the *Draft Guidelines for Non-commercial Altruistic Surrogacy using IVF as Treatment*

What follows is an explanation of the themes which underlie the *Draft Guidelines for Non-commercial Altruistic Surrogacy using IVF as Treatment*. At the time of writing, these themes take the form of a paper for further discussion and ratification by the Ethics Committee in March, 2002. It will eventually become part of the draft guidelines.

The guidelines are based on surrogacy applications submitted to the Committee by fertility clinics for ethical review, on public consultation, and on a wide range of national and international publications in the field of ethics. They apply in situations where infertility arises from medical causes.

Advances in reproductive technologies offer many potential benefits for infertile persons. However, there are also uncertainties, risks, and unknowns. These guidelines seek to balance harms and benefits in ways that respect the wishes of individuals, demonstrate caution in relation to possible harms, and give due respect to society's evolving norms. In particular, the guidelines acknowledge the intrinsic worth of persons, who must not be used as a means to an end, for example, through commercialism and commodification.

The guidelines respect the autonomy of all parties: persons seeking the assistance of others to form their families; those offering to assist; and any existing children involved in the arrangements. In this regard, choice and informed consent are essential. Respect for privacy is also important.

In particular, the guidelines seek to protect the interests of the child-to-be, including his or her need to grow up in a caring family and form a clear concept of self-identity that is underpinned with accurate information about origins.

Tikanga Maori has an important role, reflecting values evident in New Zealand society generally. Notions of caring for children in the wider group of whanau, gifting children through whangai, and knowing whakapapa underpin the guidelines. There is an emphasis on caring family relationships, and altruism within families and among close friends.

The guidelines respect the professionalism of fertility specialists and others working in reproductive medicine, counselors, and legal advisers who work with the parties in a proposed surrogacy arrangement, acknowledging their expertise and ethics as professionals. They also take into account relevant law, at the same time acknowledging that this was mostly written for purposes other than those forming the subject matter of these guidelines.

The language of the guidelines reflects graduated levels of expectation. Hence, careful attention should be given to words such as 'requires', 'expects', 'prefers', 'advises', 'must be', 'should be', 'may be', 'are likely to be'. Such language allows NECAHR on a basis of reasoned negotiation underpinned by shared ethical values, to work with applicants for ethical review, who, in turn, work with their clients.

Extract from the *Draft Guidelines for Non-commercial Altruistic Surrogacy using IVF as Treatment*

The draft guidelines take the form of aspects to be considered by the respective parties involved in a surrogacy arrangement. The main sections of the guidelines have been included below.

Providers of Fertility Services

NECAHR requires a report on the medical status of the birth mother, including the age of the mother, existing medical history, and the number of children. Information on the birth mother's age is necessary as the risks to the mother's health and likelihood of a less successful outcome increase with age. Information on the number of children is necessary in order to know whether the birth mother is likely to be capable of having a normal pregnancy. Also, a surrogate mother who already has children of her own is likely to be more aware of the medical and psychological risks to herself.

The application for ethical review should be explicit about conditions that may impact on the safety of the birth mother when undertaking treatment and pregnancy and should include documentation from medical advisers.

The treatment must be in accordance with the RTAC guidelines.

If the birth mother has a partner, the provider must discuss with the birth mother and her partner how they will ensure that they do not conceive their own child during the IVF treatment.

NECAHR considers that screening of the birth mother's partner should be the standard screening carried out for partners of women undergoing IVF treatment, i.e. for HIV and Hepatitis B and C.

NECAHR requires a provider to notify it in the case of each non-commercial altruistic surrogacy using IVF as treatment which has been approved:

when the IVF programme begins
when pregnancy is confirmed or the programme is discontinued
any adverse events
the outcome of pregnancy.

NECAHR requires that the clinic's policy take account of cultural diversity.

Commissioning parents

One or both of the commissioning parents should be the potential child's genetic parents. This means that the gametes of at least one of the commissioning parents should be used.

There should be a medical condition that precludes the commissioning mother from undertaking a pregnancy or makes pregnancy damaging to her or the child.

NECAHR prefers that the birth mother be either a family member or close friend of the commissioning parents.

Recompense may be made for expenses related to pregnancy and childbirth, but no payment should be made in lieu of employment.

Birth mother and her partner

The birth mother and her partner should have completed their family as this may reduce the likelihood that they will want to keep the child. Problems could arise if they had not completed their family or begun it, including in relation to medical complications due to the surrogacy, which then prevented further pregnancy.

If the birth mother has a partner, the birth mother and her partner must take measures to ensure that they do not conceive their own child during the IVF treatment.

Legal advisers

NECAHR requires a report from a legal adviser for each party, indicating that the party clearly understands the legal issues and the current environment in which surrogacy agreements are legally unenforceable. The same legal adviser must not advise both parties.

NECAHR does not require a formal agreement. This does not preclude a statement of intent between the parties that allows them to work through the issues and clearly state their intentions and expectations.

NECAHR advises that the parties discuss possible disputes, for example, about the custody of the child, termination of pregnancy, and life style issues during pregnancy, with their legal advisers and counsellors before the proposal is finalised. It should be noted that disputes may ultimately be resolved by a court.

Legal advisers must ensure the parties understand that the child will legally be the child of the birth mother (and her partner if there is agreement to the surrogacy arrangement), unless adopted by the commissioning parents.

Legal advisers must ensure that the parties clearly understand procedures relating to guardianship, custody and adoption and the requirements that adoptive parents have to meet, including the requirements of CYFS [Child Youth and Family Service], if they wish to adopt the child.

Counsellors

NECAHR requires counselling reports which confirm that the following issues raised by NECAHR have been discussed and, in the professional judgement of the counsellors, have been adequately understood. NECAHR prefers that two counsellors be involved, one for each family group.

Counselling must be undertaken by qualified counsellors and be culturally appropriate.

Counselling must include discussion of the following:

- the possibility of a breakdown in the arrangement such that the birth mother wishes to keep the child, or the commissioning parents do not wish to take custody of the child;
- the position of both parties in the event of a multiple birth;
- the risk of rejection of a child for any reason, e.g. if the child is born with a disability or abnormality;
- the possibility of legal termination of a pregnancy if foetal abnormality is diagnosed before birth, having regard for the Contraception, Sterilisation and Abortion Act 1977;

the possibility of the birth mother deciding against a termination in the above situation, and the subsequent care of the child;
the amount of control that genetic parents have over the birth mother's conduct of her pregnancy;
the availability of a permanent, accurate record of conception and gestation for the child;
any issues covered in a written agreement.

NECAHR expects that the parties be counselled as two separate family groups, two family groups together, and as individuals. Existing children should be included in counselling in an age-appropriate manner.

NECAHR prefers that there be a month free of counselling after the initial counselling period and then further counselling, to allow for the issues to be thought through without counselling intervention.

NECAHR expects counsellors to follow the usual counselling practice of recording the family histories of those involved in the surrogacy arrangement. If there are life experiences, for example, psychiatric problems, substance/physical/sexual abuse which may predispose any of the persons to risk when moving into a new situation, or which may pose a risk to the potential child, these must be referred to in the counsellors' reports.

A process should be set up for the resolution of disputes, for example, about the custody of the child or any other issues that arise in discussion with counsellors and legal advisers, before the proposal is finalised.

Further considerations

The Committee is prepared to consider an application deviating from the proposed guidelines. If applicants wish to deviate from any of the proposed guidelines, they should indicate this and give their reasons at the time of the application.

Please note that these guidelines which NECAHR wishes to see addressed in applications for ethical review of non-commercial, altruistic surrogacy using IVF as treatment are provisional only. NECAHR cannot at this time guarantee that the guidelines include all the issues it might wish to have addressed by applicants in such proposals. Where new issues do come to its attention, NECAHR undertakes to inform potential providers of this in as timely a fashion as possible.

The Committee welcomes comment on the proposed guidelines, to assist in the ongoing development of the guidelines. The Committee requests that previous draft guidelines be destroyed.

Ethics within and without the law

Over the past nine years, New Zealand's National Ethics Committee on Assisted Human Reproduction has opened the way for a small but significant number of persons to fulfill their wishes to form families using, because of their infertility, the technologies available to them. In the meantime, politicians, the media, watchdog groups and others have publicly commented, criticized and attempted to formulate policy and draft laws. For example, Coney and Else, in a 1999 discussion document, *Protecting Our Future The Case for Greater Regulation of Assisted Reproductive Technology*, wrote:

> No ethics committee can be expected to deal with the issues involved in surrogacy in the confused, piecemeal legislative and regulatory context in which ART is currently proceeding (p. 59).

As early as 1986, an Interdepartmental Monitoring Committee on Assisted Reproductive Technology was set up to monitor developments in the field of assisted reproductive technologies. In April 1993, a two person Ministerial Committee on Assisted Reproductive Technologies was appointed to study assisted reproductive technologies and the current situation in New Zealand and internationally, and report to the Minister of Justice. It advised in 1994 that there was already in place an adequate legal framework for surrogacy to proceed, and that the Ethics Committee should allow it to proceed (*Assisted Human Reproduction Navigating Our Future Report of the Ministerial Committee on Assisted Reproductive Technologies*, 1994).

In the meantime, in 1993, the Ministry of Health had set up the Interim National Ethics Committee on Assisted Reproductive Technologies referred to earlier in this chapter. Review of proposed surrogacy arrangements requested of a regional ethics committee, which declined to carry out the review, provided the catalyst for its establishment.

In 1996, whilst in opposition, Labour MP Diane Yates put forward a private member's Bill: Human Assisted Reproductive Technology Bill. In 1998, the National New Zealand First Coalition Government put forward the Assisted Human Reproduction Bill. Written submissions on the two Bills were called for and received in 1997/1998. In 2000, with a Labour Alliance Coalition Government in power, oral submissions on the two Bills were taken. This process was put on hold as the Government decided the Bills needed updating, given the rapid advances in technology. Also in 2000, the Government Administration Select Committee took oral submissions on the reform of the 1955 Adoption Act, currently the vehicle for commissioning parents in a surrogacy arrangement to become the legal parents of the child resulting from their gametes.

Both Bills acknowledge the limitations of legislation in this rapidly changing field, and attempt to allow for flexibility by proposing a regulatory authority of some kind. Sir Douglas Graham, then Minister of Justice, wrote about the Assisted Human Reproduction Bill in 1999:

> There are numerous difficulties in trying to produce an exhaustive list of prohibited activities. For instance, shifting social attitudes may result in some activities which are currently unacceptable gaining acceptance over time. Furthermore, there are difficulties in attempting to define for all time an exhaustive list of unethical activities in an area of rapid technological advances. Accordingly, the Government has taken the view that other activities [than those stipulated] will be dealt with through the ethical decision making structures that the Bill puts in place (p. 7).

The 1999 Bill prohibits 'the cloning of humans, the fusing of animal and human gametes, the implantation of animal or human embryos into the opposite species, and the use of human cells to develop procedures or techniques for undertaking any of those activities'. The Bill also prohibits 'the supply of human gametes and embryos for valuable consideration'. It provides for 'an information scheme intended to promote a policy of openness with respect to children born as a result of procedures involving donated gametes'. The 1996 Bill, different in its detail particularly in relation to a proposed regulatory authority, also proposes an information scheme, prohibits 'certain practices in relation to the commercialization/ sale of body parts, embryos, gametes, fetal tissue etc.', and prohibits 'improper use of reproductive technology and surrogacy'.

While these various developments have not provided the authority, guidance and sharing of responsibility that the National Ethics Committee on Assisted Human Reproduction needs and legislation should give, they have assisted the Committee to gauge public views and assess what society finds acceptable. They have helped the Committee decide how far it might push the ethical boundaries in the interests of this small, special-interest group.

The situation in November 2001 was as follows:

- Cabinet has agreed that a Special Order Paper (SOP) will be drafted which updates the Human Assisted Reproductive Technology Bill. The SOP will provide a flexible legislative framework for regulating assisted human reproductive technology and will prohibit some clearly unethical activities, for example, cloning of human beings and commercial surrogacy. The framework developed for assisted human reproduction will also be compatible with the legislative framework for the health sector generally.

- The Ministry of Justice has prepared drafting instructions in consultation with the Ministry of Health and the Ministry of Science, Research and Technology and these are currently with the Parliamentary Counsel Office (PCO) (Ministry of Health, 19 Nov., 2001).

In addition, in December 2001, a Supplementary Order Paper was introduced which will amend the Medicines Act 1981 'by placing temporary prohibitions on xenotransplantation and germ line gene therapy'. Although not intended, 'the definition of germline gene therapy may have inadvertently captured cloning'. These amendments are currently before the Finance and Expenditure Select Committee, which is due to report back to the House by 19 March 2002. The proposed restrictions to the Medicines Act will apply until 30 June 2003. This date may be extended until 30 June 2006. These extensions, it is thought, will allow adequate time to address policy and legislative matters (Ministry of Health, Feb., 2002).

This chapter began with the statement that legislation in the field of ART in New Zealand was not far away. In the meantime, New Zealand's National Ethics Committee on Assisted Human Reproduction will continue to review proposed new applications of the technologies within a framework that has been proved to work well. Considering the developments to date in policy formulation and the drafting of legislation, it may well be that ethical review will provide New Zealand's 'regulatory' framework for some time to come.

References

Assisted Human Reproduction Bill, 1998.

Assisted Human Reproduction Navigating Our Future Report of the Ministerial Committee on Assisted Reproductive Technologies (1994), Tribunals Division: Wellington.

Beauchamp, T.L. and Childress J.F. (1994), *Principles of Biomedical Ethics*, 4th edn, Oxford University Press: New York.

Coney, S. and Else, A. (1999), *Protecting Our Future The Case for Greater Regulation of Assisted Reproductive Technology*, Women's Health Action Trust: Auckland, New Zealand.

Graham, Sir Douglas (1999), 'Current Medico-Legal Issues Address to the 7th Annual New Zealand Medico-Legal Conference', *Conference Proceedings*, Institute for International Research: Wellington.

Human Assisted Reproductive Technology Bill, 1996.

Jonsen, A.R. and Toulmin, S. (1988), *The Abuse of Casuistry. A History of Moral Reasoning*, University of California Press: Berkeley.

Ministry of Health (2001), *Draft Guidelines for Non-commercial Altruistic Surrogacy using IVF as Treatment*, Ministry of Health: Wellington.

Ministry of Health (Nov., 2001), personal communication.

Ministry of Health (Feb., 2002), personal communication.

Ministry of Health (Mar., 2002), NECAHR agenda paper, Wellington.

Non-Commercial Surrogacy by Means of In Vitro Fertilization, A Report of the Interim National Ethics Committee on Assisted Reproductive Technologies (1995), Ministry of Health: Wellington.

PART II
ACCESS TO ART

Chapter 10

Access To Assisted Reproductive Technology By People With High Genetic Risks And Transmittable Diseases

Guido Pennings

Introduction

The immediate cause that gave rise to the interest in transmittable diseases is the discussion about access to infertility treatment for HIV positive people. While looking into this issue, I noticed that the more general question of providing infertility treatment to people at high genetic risk has rarely been treated explicitly. There is of course the extended debate on the acceptability of procreation in the context of genetic counseling. However, the question here is not whether we can or should actively intervene to prevent the birth of a child but whether we can or should actively participate in the creation of a child given the risks.

There are two major elements in the decision: the gravity or seriousness of the disease and the probability of its occurrence. The matter of the gravity will be discussed first. The question of what constitutes 'acceptable risk' comes second. Finally, I focus on the specific context of medically assisted reproduction and the contribution of the physicians in the procedure.

Gravity of the disease and the evaluation of the welfare of the child

The first question we have to address is when infertility treatment or medical assistance to procreation is morally acceptable. The most frequently used standard to deny infertility treatment to people is the 'best interest of the future child'. This notorious standard, however, is very difficult to define. Two difficulties arise: 1) the calculation of the life quality of a person and, in the context of procreation, of future persons; and 2) the evaluation of the quality of life presupposing that we manage to calculate this feature. If we determine that the quality of life of a future child is say 75 qaly, is 75 then sufficient since the child will have a reasonable life, or insufficient because the parents could or can have a child with 95 qaly, or too low but acceptable since the parents could not have had a child without the defect? In the literature, three standards or evaluation criteria can be found (Pennings, 1999). The strictest standard is the 'maximal welfare'. According to this standard, no medical assistance should be provided when there are indications that the life conditions of the future child will not be optimal. These indications usually are found in either characteristics of the parents, and especially of the mother, or in the social, economical and psychological circumstances in which the child will grow up. People who accept this standard argue for instance that it is best for a child to be raised by a mother and a father and that consequently lesbian couples should not receive infertility treatment. However, since there is no consensus on reliable predictive criteria for adequate parenting, it is virtually impossible to select parents in order to maximize welfare. Moreover, with such a strict standard, almost no one should have children since everyone possesses one or more characteristics that have a detrimental effect on the child.

The overwhelming majority of the authors adopt the 'minimal threshold' standard. They argue that medical assistance to reproduction is only unacceptable if the life quality of the future child is so low that it would have been better off not to have been born (Robertson, 1994; Strong and Shinfeld, 1984). This is also called the 'wrongful life' or 'worse than death' standard. According to this standard it is not wrong to bring a child into the world that would have awful prospects and opportunities (Steinbock and McClamrock, 1994). Even if parents know with 100 per cent certainty that their child will be HIV infected, it is not wrong to conceive it knowingly and intentionally since most of these children will nevertheless have a few disease-free years (Hutton, 1996). This standard expresses a minimalist view on the parental and medical responsibility for the wellbeing of the future child (De Wert, 1999).

These authors are brought to adopt this standard because they accept the 'genesis problem'. The 'genesis problem', originally formulated by Parfit, goes in three steps. Premise 1: Bringing a person into existence benefits that person. Premise 2: Without the infertility treatment, the person would not exist. Conclusion: Infertility treatment to bring a person into existence who would otherwise not have lived, only harms the person if he/she has a life not worth living.

However, the first premise is wrong: when there is no person, he/she cannot benefit from being brought into existence. The 'genesis problem' fails to distinguish the decision-making 'before birth' and 'after birth' (or, depending on one's view on the moral status of the embryo, 'before' and 'after conception'). The proponents of this argument assume a 'gate to existence', a kind of 'existential vestibule' or waiting room where the nonexisting people are waiting to be invited. Although they are placed in a pre-existence world, they are accorded the full ontological status in the sense that they already have interests that can be harmed. However, before birth, there is no person and the question is precisely whether he/she should be brought into the world (Cohen, 1996; Green, 1997). When a child is born, euthanasia or non-treatment would only be acceptable if the life of the child is not worth living. The strongest standard (substantial or devastating harm) should be used to justify this step. When we have to decide to bring a person into being, however, we should offer more than just a life worth living and we do not have to offer a perfect life. Both standards lead to highly counterintuitive judgments as they force us to condemn procreative decisions that look fully acceptable. The standard I propose is the third alternative, i.e., a 'reasonable life quality' (Pennings, 1999). A person has a reasonable life quality when he/she has the abilities and opportunities to realize those dimensions and goals that in general make human lives valuable. Deviations which cause the welfare level of the child to fall below this baseline comparison state are sufficient reasons to refuse infertility treatment. Others have proposed comparable standards. Green (1997), for instance, holds that a person should not be brought into being with 'a health status likely to result in significantly greater disability or suffering, or significantly reduced life options relative to the other children'. The rule to evaluate assisted reproductive technology is that it is morally unacceptable to bring a child into the world, and indirectly to help others to bring a child into the world, when there is a high risk that the child will not have a reasonably fulfilling life.

Probability and recklessness

The probability cannot be evaluated independently from the magnitude. There is an inverse relationship between magnitude and probability of harm which can be expressed in the following formula: the greater the gravity of the envisioned harm, the less probable it need be (Feinberg, 1984). Or, 'the greater the magnitude and probability of predicted harm, the less justifiable it is to have children' (Arras, 1990). We are interested in deciding how probable a disease which causes the child to have a life quality that falls below the threshold should be before it becomes unacceptable for parents to procreate and for physicians to assist the parents. At the start of the AIDS epidemic, the probability of vertical transmission was between 20 and 40 per cent. This risk is comparable to the 25 or 50 per cent chance of many genetic diseases. At present, when the woman is infected, the risk of vertical transmission is approximately two per cent when all precautions are taken. This probability seems insufficient to exclude HIV positive people from infertility treatment. The next question is where the cut-off lies. There is, as far as I know, no ethical theory sufficiently developed to determine precisely from which point on a risk becomes unacceptable and neither do I know of any rational argument to defend a specific cut-off.

Two important points should be considered when talking of probability in the context of genetic and transmittable diseases. Firstly, procreation always involves risk. There is no such thing as risk-free reproduction. Adopting minimal risk as a necessary condition implies either to refrain from having children at all (and not helping others) or to an absurd number of tests and checks. Secondly, the effect of the physician's collaboration on the probability changes completely depending on whether the couple is infertile or not. A number of genetic diseases like cystic fibrosis and thalassemia diminish or completely block the fertility of the affected people. When the couple is infertile, all risk for future offspring is eliminated since no future offspring will come into existence without the physician's help. Some authors have argued that in those cases it is completely incomprehensible that the physician would revive the risk (Auroux, 1999). However, the risk is revived in every infertility treatment. The base rate of major malformations in the general population is between three and five per cent. So every child created by means of infertility treatment has a chance of three to five per cent of a major illness or handicap. Moreover, if we would screen all infertility patients for the common genetic and infectious diseases in the same way as we currently do for AIDS, we will find many more infertility patients who belong to a risk group.

On the other hand, HIV positive people can have children without medical assistance. Their request for assistance is not directly motivated by the wish to have a child (since they do not need help from others) but by the wish to reduce as far as possible the risks for the partner and/or the future child. It could be argued that a physician has an obligation to promote and protect the best interests of his/her patients by taking measures that reduce as much as possible the health risks for partner and future child. This would imply that he/she should help fertile HIV positive couples to have a child in the least risky conditions. However, for fertile as well as infertile couples, he/she accepts the foreseeable two per cent transmission risk by collaborating in the parental project. Although with fertile couples he/she can argue that he/she participates with the main intention to reduce the risk, he/she remains accountable for accepting to bring into existence a certain number of infected children. However, this decision is only reckless and the doctor is only blameworthy when the risk taking is unjustified and unreasonable. According to me, this is not the case.

Positive right to infertility treatment and discrimination

A remarkable fact about the 'welfare of the child' criterion is its inconsistent application, both regarding the seriousness of the consequences for the child as regarding the probability. Several authors have pointed at the discrepancy between the accepted risk level for genetic diseases and the accepted risk for AIDS (Nolan, 1990; Rojansky and Schenker, 1995). But other comparisons should also be made. When the perinatal mortality rate of IVF twins (4.7 per cent) and of IVF triplets (8.2 per cent) is compared with the risk of HIV infection, it becomes difficult for infertility specialists to argue that the latter risk is unacceptably high (Lieberman, 1998). Moreover, multiple pregnancies are to a large extent iatrogenic which increases the physician's responsibility for the consequences.

The inconsistent use of this criterion constitutes a powerful argument for people who are denied access to assisted reproductive technology. There is a general agreement that infertile people have no positive right to medical assistance for infertility and that there is no obligation on the part of society to provide assisted reproduction to all those who request it. However, people have a generally recognized right not to be discriminated. Justice demands that like cases are treated alike. When a category of people is excluded from infertility treatment because they possess a certain characteristic (like risk for the child, reduced life expectancy, reduced competence etc.) and they are able to show that other categories of people who are accepted as patients possess the same characteristic to an equal or even a higher degree, the first group may rightly argue to be unjustly discriminated. Justice considerations would, at least for some groups, generate a positive right to infertility treatment.

A good illustration of this inconsistency is the use of 'reduced life expectancy' as a criterion. This argument has been advanced against postmenopausal women requesting infertility treatment (Pennings, 1995). Also in the context of medical assistance to HIV positive people, the chance that these children are orphaned at a relatively early age and will see their parent(s) die has been considered as a very important factor in the determination of the child's welfare. However, other categories of patients have an equally bad prognosis and are still accepted without remarks. ICSI is offered to treat infertility in male cystic fibrosis patients (Lewis-Jones et al., 2000). The reduced life expectancy may be mentioned during counseling but it is not considered as a strong contra-indication (Rodgers et al., 2000). I do not intend to pass judgment on treating cystic fibrosis patients but I want to urge for consistency in the arguments. The selective use of an argument shows prejudice and discrimination.

The patient-physician relationship

Society rarely restricts the reproductive plans of its members. Their autonomy is respected by allowing them to reproduce according to their personal moral principles. Non-directive counseling in the field of genetics is the prime illustration of this right. However, while parents have the right to take certain risks without interference from others, this does not mean that doctors should assist couples in taking these risks. The explanation lies in the fact that the parental liberty has little to do with the moral acceptability of their decision. The ultimate reason for not coercing parents lies in the disproportional violation of their autonomy and body integrity implied by forced sterilization or abortion. Even if we all agree that a certain parental decision is morally unacceptable, this does not automatically entail that we accept forced intervention as a means to prevent it.

The matter of the responsibility of the fertility specialist has received remarkably little attention. This question cannot be solved if we do not manage to determine what precisely his/her role is in the process and more specifically his/her position in relation to the intentional parents. I consider the intentional parent(s) as the principal(s). The principals are those who initiate and support the project. The other persons involved, like gamete donors, surrogate mothers, and medical personnel are all collaborators to this project. This structure generates a possible conflict: on the one hand, the physician makes a judgment on what the parents can or should do and the risks they are allowed to take. On the other hand, he/she contributes to the outcome of the project and thus becomes accountable for the result. So even if he/she thinks that a certain risk taking is acceptable for the patients, he/she may consider this risk as too high to justify his contribution. Building on the general discussion regarding the moral (and legal) responsibility of complicity, a collaborator is morally responsible for the consequences of a project if he/she intends to bring them about or if he/she is able to foresee that his/her contribution will have a high chance of bringing them about. Since the physician freely participates in the parental project, he/she is (partly) responsible for the welfare of the children who result from this. When a child is born with a handicap or illness, he/she is co-responsible as a collaborator. However, responsibility does not imply blameworthiness or culpability. To determine culpability we have to analyze his/her actions and decisions. As I mentioned earlier, risk-taking is only condemnable if the person, fully aware of the risks of bringing about possible consequences, goes ahead even when reasonable people consider the risk as too high (Smith, 1989).

The main distinction is between acceptability of assistance and obligation of assistance. Physicians should be allowed to help in establishing a pregnancy in all cases in which procreation by the parents is deemed acceptable. The physician as a collaborator in the parental project is not bound by a higher standard than the principals. If the decision by the principals is morally acceptable, helping the principals is also acceptable. The physician can offer several reasons for helping the couple: he/she may share the same values, principles and views on responsible parenthood; he/she may act out of respect for the autonomy of the parents in those cases when he/she considers the decision as 'wrong but morally permissible'; he/she may argue that his/her participation is essential for the best interests of the couple and/or the offspring. The latter justification applies when his/her participation reduces the risk for the future child in fertile couples.

It has been argued that physicians are bound by higher standards than the parents. According to the deontological code of their profession, they would have a commitment to try to maximize wellbeing (Savulescu, 1999). Applied to the level of risk, they should accept less risk than those who do not occupy this role. The discrepancy could be explained by shifting the focus of the responsibility of the doctor. Instead of seeing the physician mainly as a helper of the intentional parents, the physician is held directly responsible for the well-being of the future child. However, why do the obligations of the parents towards the person they intend to bring into existence differ from the obligations of those who assist the parents in the creation? Both parties should respect the reasonable welfare standard. The fact that the parents will take care of the child when it is born with a serious handicap is irrelevant. This intention does not make it acceptable to create a handicapped child. Both the parents and the fertility specialist should refrain from procreation if there is a high risk that the child will be harmed.

However, although the doctor should be free to participate, he/she has no moral duty to help in all cases. The helper is morally responsible if he/she intends to help the principals to realize their plan (i.e., to have a child) and/or if he/she contributes while knowing what the principals intend to do. The physician cannot hide behind the parents and claim that they carry full responsibility for the foreseeable consequences. Precisely this moral responsibility as a consequence of the causal and intentional contribution gives the physician the right to appeal to conscientious objections. The physician is not obliged to help the patients when they request an option that falls below the personal standard of the physician. Participation in a project he/she considers morally impermissible would render him morally blameworthy.

Gamete donation as an alternative

A final point is whether gamete donation can be imposed on patients as a means of reducing or eliminating the risk for the offspring. It could be argued that if a couple wants the doctors to collaborate, the doctors have the right (or obligation) to impose certain restrictions in order to protect the welfare of the future child. The National Advisory Ethics Committee in France recommended that, in order to eliminate the risks, adoption and donor insemination should be the preferred modes of treatment of HIV positive men (Comité consultatif national d'éthique pour les sciences de la vie et de la santé, 1998). This was remarkable since the French CECOS Genetics Advisory Board only allows the use of donor sperm as an alternative when the genetic condition is serious, the risk of transmission is very high and the illness is not detectable prenatally (Novaes, 1999). Again, the ready proposal of donor material as a solution demonstrates the different attitude towards risk taking by HIV positive people and by people at high genetic risk.

This matter is important since it is considered morally relevant whether the parents have any alternatives. The question raised here is which options are considered as real alternatives. Harris (1992) for instance argued that 'if the particular parents must have disabled children if they are to have children at all, then they will be blameworthy only of the children would be wronged by existence'. Both for AIDS and genetic diseases that are not detectable prenatally, the risk for the offspring exists for the rest of the parents' life and there is no way they will be able to have healthy children without risk taking. However, the situation can only be labeled as a 'Different Number Choice' (one child with the risk of a handicap or no child) if the use of donor gametes is excluded beforehand. The couple could have healthy children if they drop the condition of a full genetic link. The risk for the child is balanced against the interest in genetic affinity of the parents. How high should the risk be before the use of donor gametes should be imposed? Savulescu (1999) discussed the case where a couple of which both partners were affected by cystic fibrosis requests IVF. The child would be certain to have the disease. Given the seriousness of the disease and the probability of 100 per cent, I agree that it is unacceptable to assist the parents in this project. Still, I believe that the instrumental value of the genetic relatedness outweighs a much smaller risk like the risks of transmission when the male partner is seropositive. Demanding the use of donor sperm for such a small risk requires a complete denial of the value of the genetic link. Nevertheless, as the discussion on the acceptability of ICSI for male patients with chromosomal aberrations shows, a broad debate on the balance between genetic relatedness on the one hand and health risk for the offspring on the other hand is very much needed. The increasing possibilities offered by the techniques of medical assisted reproduction will inevitable confront us with the question of where to stop as well in terms of investment and effort as in terms of risks for all people involved.

Conclusion

As a general rule, access to infertility treatment should be refused when there is a high risk of a serious disease in the offspring. A serious disease is a handicap, illness or disability which does not permit the child to have a reasonable life quality. If the risk is evaluated as too high, the use of donor gametes can be imposed as a condition for medical assistance.

The physician has the right to collaborate in the realization of the parental project if the parental project itself is considered morally permissible. Due to his/her causal and intentional contribution, he/she inevitably shares the responsibility for the outcome of his/her actions. As a consequence, the doctor is free to refuse participation when he/she considers the wish for a child as morally unacceptable. However, since procreation is considered a major part of most people's life plan, he/she should be able to justify his/her refusal. The reasons he/she provides for excluding some categories of persons should be consistently applied to all relevantly similar cases.

References

Arras, J.D. (1990), 'AIDS and reproductive decisions: having children in fear and trembling', *Milbank Quarterly*, Vol. 68, pp. 353-82.

Auroux, M. (1996), 'Faut-il traiter l'infertilité chez les couples dont l'un des partenaires est séropositif?', *Contraception Fertilité Sexuologie*, Vol. 27, pp. 118-20.

Cohen, C.B. (1996) '"Give me children or I shall die!" New reproductive technologies and harm to children', *Hastings Center Report*, Vol. 26, pp. 19-27.

Comité Consultatif National d'éthique pour les Sciences de la Vie et de la Santé (1998), 'Avis et recommendations sur les "Problèmes éthiques posés par le désir d'enfant chez les couples ou l'homme est séropositive et la femme séronégative"', *Les Cahiers du C.C.N.E.*, Vol. 15, pp. 3-10.

De Wert, G. (1999), *Met het oog op de toekomst*, Thela Thesis, Amsterdam.

Feinberg, J. (1984), *The moral limits of the criminal law. Vol. 1. Harm to others*, Oxford University Press, New York, Oxford.

Green, R.M. (1997), 'Parental autonomy and the obligation not to harm one's child genetically'. *Journal of Law, Medicine & Ethics*, Vol. 25, pp. 5-15.

Harris, J. (1990), 'Wrongful birth', in D.R. Bromham, M.E. Dalton and J.C. Jackson (eds.), *Philosophical ethics in reproductive medicine*, Manchester University Press, Manchester, New York, pp. 156-170.

Harris, J. (1992), *Wonderwoman and superman: the ethics of human biotechnology*, Oxford University Press, Oxford, New York.

Hutton, N. (1996), 'Health prospects for children born to HIV-infected women', in R.R. Faden and N.E. Kass (eds.), *HIV, AIDS and childbearing: public policy, private lives*, Oxford University Press, Oxford, New York, pp. 63-77.

Lewis-Jones, D.I., Gazvani, M.R. and Mountford, R. (2000), 'Cystic fibrosis in infertility: screening before assisted reproduction', *Human Reproduction*, Vol. 15, pp. 2415-7.

Lieberman, B. (1998), 'An embryo too many?', *Human Reproduction*, Vol. 13, pp. 2664-6.

Nolan, K. (1990), 'Human immunodeficiency virus infection, women, and pregnancy', *Obstetrics and Gynecologic Clinics of North America*, Vol. 17, pp. 651-68.

Novaes, S.B. (1999), 'Making decisions about someone else's offspring. Geneticists and reproductive technology', in M. Fortun and E. Mendelsohn (eds.), *The practices of human genetics*, Kluwer Academic, Dordrecht, Boston, London, pp. 169-84.

Pennings, G. (1995), 'Age and assisted reproduction', *Medicine and Law*, Vol. 14, pp. 531-41.

Pennings, G. (1999), 'Measuring the welfare of the child: in search of the appropriate evaluation principle', *Human Reproduction*, Vol. 14, pp. 1146-50.

Robertson, J.A. (1994), *Children of choice*, Princeton University Press, Princeton, New Jersey.

Rodgers, H.C., Knox, A.J., Toplis, P.J. and Thornton, S.J. (2000), 'Successful pregnancy and birth after IVF in a woman with cystic fibrosis', *Human Reproduction*, Vol. 15, pp. 2152-3.

Rojansky, N. and Schenker, J.G. (1995), 'Ethical aspects of assisted reproduction in AIDS patients', *Journal of Assisted Reproduction and Genetics*, Vol. 12, pp. 537-42.

Savulescu, J. (1999), 'Should doctors intentionally do less than the best?', *Journal of Medical Ethics*, Vol. 25, pp. 121-6.

Smith, P. (1989) 'Recklessness, omission, and responsibility: some reflections on the moral significance of causation', *Southern Journal of Philosophy*, Vol. 27, pp. 569-83.

Steinbock, B. and McClamrock, R. (1994), 'When is birth unfair to the child?', *Hastings Center Report*, Vol. 24, pp. 15-21.

Strong, C. and Shinfeld, J.S. (1984), 'The single woman and artificial insemination by donor', *Journal of Reproductive Medicine*, Vol. 29, pp. 293-9.

Chapter 11

Should There Be Limits On Who May Access Assisted Reproductive Services? A Legal Perspective

Kristen Walker

Introduction

The question of access to assisted reproductive services has been the subject of intense public, media and political debate in Australia following Dr McBain's successful case against the State of Victoria in 2000.[1] In the *McBain* case, Justice Sundberg of the Federal Court ruled that parts of the *Infertility Treatment Act* 1995 (Vic) that required a woman to be married or in a de facto relationship with a man were invalid because they discriminated against single women[2] and were thus inconsistent with section 22 of the Commonwealth *Sex Discrimination Act* 1984.[3] Although not the first court to have ruled on the issue,[4] Sundberg J's decision caused an outcry from churches, "family values" organizations and coalition politicians. Prime Minister Howard declared within days that the decision would be overruled by legislation,[5] and the Catholic Church sought review of the decision in the High Court.[6]

In this paper I will consider whether regulation of access to assisted reproductive services is appropriate and, if so, what form that regulation should take. There have been many inquiries into assisted reproductive services over the years - one in almost every Australian State, as well as in other jurisdictions - but some of these have not considered questions of access at all,[7] while others have simply assumed that imposing a requirement that those accessing services be married[8] or in a heterosexual de facto relationship[9] will safeguard the interests of any child born.[10] I will argue that some form of regulation of access is appropriate, but that regulation should operate in a non-discriminatory fashion, to protect the interests of any children born through assisted reproductive services and not to enforce a particular family form.

Should there be regulation?

Under a liberal theory of law, legal prohibitions on individual action are justified to prevent harm to others, but not otherwise. As Mill put it in 1859:

> [T]he only purpose for which power can be rightfully exercised over any member of a civilised community, against his will, is to prevent harm to others. His own good, either physical or moral, is not a sufficient warrant. He cannot rightfully be compelled to do or forbear because it will be better for him to do so, because it will make him happier, because, in the opinion of others it would be wise, or even right (Acton, 1972, p.78).

This approach to power seeks to maximize individual autonomy, though not at the expense of others' autonomy. Inherent in this formulation is a principle of equality - the right to liberty or autonomy is not only a right for some, it is a right for all. That is, there is an *equal* right to autonomy, subject always to the harm principle. The importance of equality in liberal theory cannot be overlooked - indeed, for an egalitarian liberal, where there is a conflict between liberty and equality, liberty must lose (Dworkin, 1987).

Mill's harm principle applies not only in relation to freedom of thought, expression and conscience, but also in relation to our actions. Quoting Mill again,

> the principle requires liberty of tastes and pursuits; of framing the plan of our life; of doing as we like, subject to such consequences as may follow: without impediment from our fellow creatures, so long as what we do does not harm them, even though they should think our conduct foolish, perverse, or wrong (Acton, 1972, p.81).

Building on Mill, contemporary scholars such as Dworkin (1993, pp.166-7) have articulated the proposition that individuals have the moral right and responsibility to "confront the most fundamental questions about the meaning and value of their own lives for themselves, answering to their own consciences and convictions". Dworkin has used this to fashion an argument for strong individual rights that can operate to trump certain collective decisions. And Hart (1986) used a Millian approach in his famous debate with Devlin concerning the relationship between law and morality in the context of the partial decriminalization of homosexuality.

Mill's and Dworkin's work has already been applied in the context of assisted reproductive services. For example, Harris (1998), building on

Dworkin's work, has articulated a right of procreative autonomy that includes access to assisted reproductive services. This right can only be overridden where there is a compelling reason for denying a person control over her or his reproductive decisions. A compelling reason would require more than simply that "a majority found the ideas disturbing or even disgusting" (Harris, 1998, p.36). In a similar fashion, Robertson (1994) has argued for a right to procreative liberty, a negative right precluding state interference with reproductive choices. These approaches each recognise that decisions about procreation are "central to personal identity, to dignity and to the meaning of one's life" (Robertson, 1994, p.24).

On a liberal approach, then, in order to justify restrictions on access to assisted reproductive services, which reduce the autonomy of women, doctors and in some cases sperm donors, it must be demonstrated that there is a compelling reason for such restrictions: they must prevent harm - possibly serious harm - to others. In relation to the present restrictions, we must ask who is harmed by lesbians or single women accessing assisted reproductive services? There are several persons to be considered: the woman involved; the sperm or egg donor; the families of the woman and the donor; the doctor; other members of the Australian community; and any child that may be born as a result of any procedure.

The woman seeking access and the sperm or egg donor

In so far as the woman who wishes to access services is concerned, I argue that the woman involved is the best judge of what might harm her and whether to bear that harm. She has chosen to avail herself of the service in question and, although this may involve invasive medical procedures, she has taken the view that the potential benefits of the treatment outweigh any harm. An assessment of harm to the individual must generally be a decision for that individual to make - in this case, the decision of the woman, on the advice of her doctor, not the decision of the state or of a church. As Mill observed, "[she] is the person most interested in [her] own wellbeing: the interest which any other person, except in cases of strong personal attachment, can have in it, is trifling, compared with that which [she herself] has" (Acton, 1972, p.144). Although there are clear cases where the state can protect a person from self-inflicted harm, these are generally extreme cases involving incapacity or mental illness. An application of the harm principle that allows the State to regulate *anything* an individual does that the State considers may harm that individual effectively negates the operation of the harm principle and unacceptably reduces individual autonomy.

Furthermore, excluding certain women from access to medically assisted reproductive services may in fact cause harm to the woman involved, as she is denied access to safe, screened sperm. As a result, many women locate a donor though informal networks and arrangements, or engage in unprotected heterosexual sex. This means they use sperm that has not been screened and quarantined and are placed at risk for sexually transmitted diseases such as HIV or hepatitis. This is one of the key reasons why many doctors support open access to assisted reproductive services[11] - to minimize harm to the woman involved and to any child born as a result, which will be discussed below. In addition, some women are prevented by existing legislation from having a child at all, it they do not have another way to obtain sperm or sufficient resources to travel interstate - this also causes harm to the woman concerned, in that she is denied the positive outcomes of having and raising a child.

Arguments based on individual autonomy apply also to the donor of eggs or sperm - she or he has chosen to donate eggs or sperm for use by a clinic or by a particular woman or couple. The donor has made an informed choice[12] about the consequences of that decision for her- or himself, which may include parenting a child, not parenting a child with whom she or he has no contact, or not parenting a child and being contacted by that child later in life (depending on the State in which the treatment occurs - this issue will be discussed further below). Some of these consequences may be distressing and thus harmful to the donor - but the decision to donate, in so far as it concerns harm to her- or himself, is one for the donor to make.

The families of those seeking access

The families of those seeking to access assisted reproductive services also need to be considered in determining the harm question. It may be in some circumstances that access to assisted reproductive services will not be in the best interests of a person's existing children or other family members. However, it cannot be said that allowing single women and lesbians access to assisted reproductive services will always involve harm to their families - on the contrary, in many cases it is likely to bring great joy and benefit to those families. Thus this does not provide a basis for regulating access to assisted reproductive services by discriminatory exclusions.

Harm to the doctor

It is difficult to see what harm may come to a doctor by permitting lesbians and single women access to assisted reproductive services - indeed, the doctor's autonomy is improved, as he or she no longer has the threat of prosecution and punishment present for permitting access. However, there is one way in which some doctors may be harmed by removing restrictions on access: this may result in a doctor who wishes to refuse to treat a single woman or a lesbian being sued for unlawful discrimination. However, I contend that this kind of harm - a restriction on one's ability to discriminate - is not sufficient to justify regulation, as discrimination itself causes harm and is contrary to the fundamental principle of equality that underpins both domestic and international legal arrangements. In providing services of any kind, all persons are subject to the various anti-discrimination laws that Parliament has enacted. These laws place limits on individual autonomy in order to prevent the harm of discrimination, and the exposure of doctors to such laws is not a justification for continuing the discriminatory restrictions on access to assisted reproductive services. This is linked to the issue of conscientious objection, which I shall discuss in Part V, below.

Harm to other members of the Australian community

There are no doubt many amongst us who feel profoundly that lesbians and single women ought not be able to access assisted reproductive services. Such feelings may be based on religion, they may be based on concern for any child born, they may be based on a general concern about the breakdown of traditional morality or the decline of marriage. Waldron (1993, p.116) has described such feelings as "moral distress - the fact that someone is distressed because of what he takes to be the immorality or the depravity of another's behaviour". However, a strongly held feeling about how another person should behave - even distress that may be experienced as harm - is insufficient to justify regulation. When Mill and others speak of "harm", they do not mean offence or harm to feelings (Acton, 1972; Dworkin, 1984; Waldron 1993). Indeed, Mill makes this very clear:

> There are many who consider as an injury to themselves any conduct which they have a distaste for, and resent it as an outrage to their feelings; ... But there is no parity between the feeling of a person for his own opinion, and the feeling of another who is offended at his holding it; no more than between the desire of a thief to take a purse and the desire of the right owner to keep it (Acton, 1972, p.152).

If harm were interpreted this broadly, it would cease to be a meaningful limit on state power to regulate, as the views of the majority - or even a minority - would always prevail over the individual's interests (Waldron, 1993, p.125). We as a society have accepted that this is so in choosing, for example, to decriminalize various sexual activities of which a significant number of people disapprove, such as prostitution, pornography, adultery or sexual conduct between persons of the same sex. In the context of access to assisted reproductive services, the fact that some members, even a majority, of the Australian community are offended and thus "harmed" by permitting access to assisted reproductive services by lesbians and single women is not sufficient to justify regulation - unless their concerns about *actual harm* to the child are correct. I will thus now examine the contention that such access would harm any child born as a result.

Harm to the child born through assisted reproductive services

The central argument made by those who oppose non-discriminatory access to assisted reproductive services is that the children who may be born to lesbians and single women will suffer harm. However, there is no evidence that this is so. A mere assertion of harm cannot justify regulation, the harm must be demonstrable - the burden is on those who seek to impose restrictions on liberty (Robertson, 1994, p.24). I have explained elsewhere (Walker, 2000) that the argument based on the welfare of the child reflects a series of inaccurate and stereotyped assumptions: that every child needs two parents living in the home; that every child needs a father living in the home; and that lesbians will detrimentally influence the gender or sexual identity of their children. The available evidence in fact suggests that none of these propositions is correct - that there is no harm to children born to single women or lesbian couples using assisted reproductive services (Walker, 2000; Golombok, 2000; Parks, 1998; Burghes, 1996).

Furthermore, in the debates around access to assisted reproductive services there has been little attention paid to the fact that two-parent heterosexual families may provide an insecure environment for children. Indeed, some two-parent families provide an unhealthy or even dangerous environment for children's development (O'Neill, 1994). Yet most legislation does not exclude *any* infertile heterosexual couples from accessing assisted reproductive services, regardless of indicators that they may not be appropriate parents. In the Queensland anti-discrimination case mentioned above, for example, the doctor who refused to provide treatment to a lesbian woman acknowledged that:

> Even if the doctors at [the clinic] thought people were completely unsuited to being parents, and even if they had been through psychological testing which had confirmed their unsuitability, they would still probably receive treatment so long as they were married or in a stable heterosexual relationship.[13]

Thus the exclusion of lesbians and single women from access to assisted reproductive services seems not so much to be about concern for the welfare of the child as it is about protecting a particular family form.

I argue that we should have regard to the interests of children born through assisted reproductive services. It is possible that allowing some people access to assisted reproductive services may result in harm to a child born as a result. There is, however, no evidence that the interests of children will be harmed by permitting lesbian couples and single women to access assisted reproductive services. Thus if regulation is to occur, it should not discriminate on the basis of sexual orientation, marital status or any other basis that is irrelevant to a person's ability to care for a child, such as "race, colour, ... language, religion, political or other opinion, national or social origin, property, birth or other status".[14] Thus rather than examining the social science evidence on the question of single mother and lesbian parenting, I want to consider a different question - if we rule out discriminatory regulation of access to assisted reproductive services, are we left with no regulation, or are there reasons to have a regulatory system based on criteria other than sexuality or marital status?

I think that there are people in relation to whom we may have a justifiable concern that they pose a risk of harm to a child - for example, a person with a history of child abuse. The question then is whether the *possibility* - or perhaps the *probability* - of harm justifies regulation and, if so, what kind of regulation is appropriate. It would be possible, of course, simply to leave the matter in the hands of the doctors who provide the services and the counselors who, in Victoria, are required to provide counseling as a matter of law.[15] However, it is incredibly difficult to know what sort of parent any given individual will make in the future, and to leave such a decision unregulated would allow too much scope for subjective value judgements, in my view. If people are to be excluded, this should not simply be at the discretion of the doctor - there should be legislative guidance.

Given the difficulty of assessing a person's future parenting abilities, it is certainly arguable that there should be no regulation - just as there is no regulation of procreation by penetrative heterosexual intercourse. Smart (1989, p.106), a prominent feminist legal theorist, has voiced concerns about using law to regulate assisted reproductive services, noting that "it is impossible to ensure that legislation, once in force, will be used progressively in the future". Yet to many people, a complete absence of regulation seems inappropriate - should we really assist a person or couple to have a child if we are persuaded that their child will be at risk of harm? But even this question leaves open the question of the degree of risk we foresee (which is merely informed speculation) and the seriousness of the harm that may result. Do we require that harm be more likely or not? Or is a slight risk sufficient? Is fear of teasing sufficient harm? Or do we mean serious harm - physical or mental abuse of some kind? And who will evaluate these risks, using what criteria? Is it perhaps better to allow access to those who want or need assisted reproductive services and then to allow the usual legal controls over child abuse and neglect to take their course once a child is born?

The form of regulation

If some form of regulation of access to assisted reproductive services is to be adopted, the question is then what form that regulation should take. It would be possible, for example, to regulate access to assisted reproductive services in the same way that adoption is currently regulated (though without the discriminatory aspects of the current adoption regimes).[16] However, adoption regimes in Australia are complex, involving invasive and time-consuming examination of the family involved and ultimately a court order to formalize the adoption. This regime is appropriate in cases where a child is already in existence, where the interests of the birth parents, the child and the adoptive parents all need to be considered and where an adoption order terminates the birth parents' existing parental rights and duties. It is not appropriate, I suggest, for assisted reproductive services, where there is no child in existence and no birth parents who relinquish their rights.

A less stringent approach would be possible and, I argue, more appropriate for assisted reproductive services. There are two possible models: first, access would only be granted if the welfare of the child has been taken into account. Second, a presumption of access could be adopted, but certain limited classes of persons would be excluded. I will consider each of these in turn.

Model 1: Welfare of the child

The first model I want to examine is the model based on the welfare of any child resulting from assisted conception. This model leaves the decision about whether to provide assisted reproductive services to a person or couple primarily in the hands of the doctor, though it may include the possibility of review by an expert body or by the courts, or both. It is this model that has, in effect, been adopted in the United Kingdom under the *Human Fertilisation and Embryology Act* 1990 (UK). Section 13(5) of that Act provides that:

> A woman shall not be provided with treatment services unless account has been taken of the welfare of any child who may be born as a result of the treatment (including the need of that child for a father), and of any other child who may be affected by the birth.

Notably, this provision does not make the welfare of the child who may be born the paramount or overriding consideration - doctors must simply take into account the welfare of the child as one factor amongst several.[17] In addition, the UK Act adds as one particular aspect of the welfare of the child the "need of [the] child for a father", a "need" that is not necessarily borne out by research on the welfare of children, discussed above. In my view, if one were to adopt the welfare of the child model of regulation, it would be preferable simply to refer to the welfare of the child, without this additional rider.

This approach to regulation of access has a certain appeal - it seems intuitively right that we should consider the welfare of the child who may be born through assisted conception. Yet I suggest that on closer examination there are various difficulties with this approach to regulation. The first, and perhaps the greatest, is that the standard adopted - the "welfare of the child" - is vague and subjective (Mumford et al, 1998; Blyth & Cameron, 1998). This standard may result in prejudice being dressed up as a concern for the welfare of the child - for example, in the exclusion of lesbians, transgenders or disabled persons from access to assisted reproductive services. And what one person views as being in the best interests of a child - such as religious education in a particular faith - may be profoundly repugnant to another person. Mumford et al have pointed out that "even if one attempts an objective assessment of welfare, it is difficult for anyone to say with justified confidence which of a number of choices is in a child's best interests" (Mumford et al, 1998, p. 2353). Even Baroness Warnock, the head of the Warnock Committee whose report provided the basis for the UK legislation, acknowledged that regulation may involve "guessing about the good of the child".[18] And as Mnookin (1975, p.260) has observed, determining what is in the "best interests" of a child "poses a question no less ultimate than the purposes or value of life itself".

It is true that courts have, in family law, been applying this standard over many years - but judicial practice is based on extensive and time-consuming investigation of the parties' situations by social workers and, often, evidence from the child. An investigative process would be cumbersome in the assisted reproductive services context and, I have argued, inappropriate, as there is, of course, no child to be observed or to give its views. Further, Pennings (1999) has suggested that there are at least three ways of interpreting the welfare of the child principle: a "maximum welfare" approach, where "one should not knowingly bring a child into the world in less than ideal circumstances"; a "minimum threshold" approach, where "we should only condemn procreation in those circumstances where the expected welfare of the child would be very low"; and the "reasonable welfare" approach, that mediates between these two extremes to propose a standard of what is "acceptable or reasonable". Thus a threshold issue if we are to use the welfare of the child approach is the choice of approach. Pennings advocates a "reasonable welfare" approach, but I suspect this perpetuates the subjectivity of the welfare standard.

In the UK, the HFEA has provided some additional guidance to doctors and clinics in its Code of Practice that goes some way to remedying the vagueness of the standard, but nonetheless leaves room for subjective value judgments. The Code states that those seeking treatment "are entitled to a fair and unprejudiced assessment of their situation and needs". It also provides a list of factors to be considered in relation to those seeking treatment:

(a) their commitment to having and bringing up a child or children;

(b) their ability to provide a stable and supportive environment for any child produced as a result of treatment;

(c) their medical histories and the medical histories of their families;

(d) their health and consequent future ability to look after or provide for a child's needs;

(e) their ages and likely future ability to look after or provide for a child's needs;

(f) their ability to meet the needs of any child or children who may be born as a result of treatment, including the implications of any possible multiple births;

(g) any risk of harm to the child or children who may be born, including the risk of inherited disorders or transmissible diseases, problems during pregnancy and of neglect or abuse; and

(h) the effect of a new baby or babies upon any existing child of the family.[19]

If the use of donated gametes is involved, then additional factors must be considered:

(a) a child's potential need to know about their origins and whether or not the prospective parents are prepared for the questions which may arise while the child is growing up;

(b) the possible attitudes of other members of the family towards the child, and towards their status in the family;

(c) the implications for the welfare of the child if the person providing the gametes for donation is personally known within the child's family and social circle; and

(d) any possibility known to the centre of a dispute about the legal fatherhood of the child.[20]

Further guidance is also provided in relation to the provision of services in circumstances where there will be no legal father:

(a) … Centres are required to have regard to the child's need for a father and should pay particular attention to the prospective mother's ability to meet the child's needs throughout their childhood. Where appropriate, centres should consider particularly whether there is anyone else within the prospective mother's family and social circle willing and able to share the responsibility for meeting those needs, and for bringing up, maintaining and caring for the child.[21]

These factors provide a significant fleshing out of the welfare of the child concept, but most of them will be incredibly difficult to assess, especially if the person(s) seeking treatment do not already have children. Many of them could involve no more than speculation. And several of them may lead to discrimination on the basis of age or disability.

This approach also has implications for the cost and timing of access to services. To truly consider a factor such as a person's ability to "provide a stable and supportive environment" would require quite invasive assessment techniques. If in fact applied, and there is a suggestion that, in the UK, the standard is ignored by clinics (Lee & Morgan, 2001, p.167), the welfare of the child approach would require an assessment process similar to that used in the context of adoption - this would be time-consuming and costly and, I argue, inappropriate.

Further, it is by no means clear that doctors are the appropriate persons to assess the welfare of the child (Daniels & Taylor, 1993, pp.1478-9). Indeed, Mumford et al (1998, p.2353) note that "there is residual unease amongst doctors as to how their duties are to be fulfilled". If this standard is to be adopted, then it will become essential to involve social workers or counselors in the decision-making process. Others have suggested an "interdisciplinary selection panel" (Daniels & Taylor, 1993, p.1479). In addition, I argue that some form of review is essential so that doctors are not simply left with unfettered discretionary power. Review could be provided initially by a licensing body such as the HFEA in the UK or the ITA in Victoria, with an additional possibility of judicial review.[22] Doctors should also have a duty to record their reasons for permitting or denying access to assisted reproductive services and to include such information in annual reports to the licensing authority. Over all, however, I argue that the problems associated with the welfare of the child standard outweigh its benefits.

Model 2: Presumption of access with limited exclusions

In this model, the basic rule would be that any woman or couple seeking assisted reproductive services would be eligible for treatment, unless they fell into one of several possible categories for exclusion (discussed below). The decision would be one initially for the doctor, acting on the advice of a counselor or social worker, but with an avenue for review of the doctor's decision. Thus a person or couple denied access should be able to seek review by an appropriate tribunal, and then by the courts, so that the decision is not solely that of the doctor. Again, there should be a requirement that doctors record their reasons for permitting or denying access and include such information in annual reports. This mechanism would give greater certainty to those making decisions about access, and may achieve a better balance between the equality and autonomy of the woman or couple and the interests of any child born as a result of treatment. Essentially this approach is based on a minimum threshold approach that in the absence of general community consensus about what makes a good parent, we should adopt a cautious approach that ensures the basic safety of the child but does not involve value judgments about parenting abilities (Pennings, 1999; Jansen, 1997).

The question then becomes what grounds should be the basis for exclusion. This is a difficult question on which there will be much disagreement, but some possible grounds for exclusion include:

- persons with a conviction for any form of child abuse;
- persons over a certain age who may well become old or infirm while their child requires care;

- persons addicted to alcohol or drugs;
- persons with a serious mental illness that would preclude them from providing appropriate care for the child; and
- persons with a serious disability that would preclude them from providing appropriate care for the child.[23]

Some of these are clearly related to the criteria set out by the HFEA in its Code of Practice. But rather than simply requiring a vague and subjective assessment of the welfare of the child, a doctor under this model would need to be satisfied that one of the above problems existed in order to exclude a person from treatment. I acknowledge, however, that in relation to some suggested bases for exclusion, a subjective assessment would be involved - absolute certainty is difficult to achieve outside of the criminal conviction category.

Interestingly, in the US where there is no real regulation of access to assisted reproductive services, research indicates that 98% of doctors and clinics excluded persons with a conviction for child abuse, 40% excluded those suspected of child abuse, 60% excluded those whose consumption of alcohol is excessive, 45% excluded those who used marijuana, and 30% excluded those with a history of schizophrenia. Many, however, were undecided about these issues (Stern et al, 2001, p.592). Thus there is evidence of considerable disagreement amongst practitioners on exclusion of various categories of person, with the exception of those convicted of child abuse. Legislative guidance is thus appropriate, I suggest.

The above list provides some suggestions for grounds of exclusion. Yet it must be acknowledged that each of these categories raises difficult issues. If we exclude persons with a mental illness or a disability, are we simply bringing in a different form of unjustifiable discrimination? Are we engaging in eugenics? Does it make a difference if the person seeking to have the child has a partner or relative who would assist in caring for the child? If we exclude those addicted to drugs and alcohol, how do we measure this - what constitutes addiction? Do we include nicotine addiction? Do we exclude only those who are addicted to illegal drugs, or do we exclude those addicted to prescription drugs such as valium? If we are concerned about older people who may not be able to care for their child, do we also morally condemn the young person with cancer who becomes pregnant so that she may give birth before she dies? And if not - if we react instead with compassion for her position - why is her position different from that of the older person? Do we fall back on arguments based on "natural" child-bearing age - and if so, how can we justify assisted reproductive services at all, given that, by definition, it assists those who cannot conceive "naturally"? And what of the fact that men of 70, 80 or even 90 father children in the ordinary fashion, without condemnation? Again, does it make a difference whether the person has a partner or not?

Even the person with a conviction for child abuse, a seemingly straightforward category of exclusion, may raise thorny issues. Do we exclude the 40 year old man who, when aged 19, had a sexual relationship with a 15 year old girl, and who has no other convictions? What of the 15 year old who was convicted of molesting an eight year old, but 20 years on has no further convictions? Or should we only be concerned with convictions in the last 5 years? Or 10 years? Do we exclude the young gay man who was convicted of an offence under discriminatory age of consent laws? Or the mother who, during a period of post-natal depression, harmed her child? And, of course, merely acting on convictions could allow those who have been charged and acquitted, or never charged, to slip through. Does the fact that our system of regulation will be imperfect mean that we should have no system at all? Further, the number of persons seeking access to assisted reproductive services who fall into these categories will no doubt be incredibly small - so do we need regulation to deal with such rare problems?

Some might also suggest other grounds for exclusion. For example, some regimes prohibit the use of sperm removed from a dead man or an embryo formed from the gametes of a person who later dies.[24] And in Britain, a woman who had a conviction for prostitution was denied access to services.[25] Others might suggest that any conviction for an offence of violence should result in exclusion - as is the case in South Australia.[26] HIV status might also be suggested as a basis for exclusion (Lee & Morgan, 2001, p.163), though in some jurisdictions it is recognised that assisted reproductive services may provide a safe way for people living with HIV to have children.[27] And others might object to providing services to a woman who wishes to become pregnant only to abort the fetus to provide tissue for a therapeutic operation to save her existing daughter's life (Lee & Morgan, 2001, p.163).

I do not purport to have the answers to these questions - in fact I consider them to be incredibly difficult. But if we wish to have a system that excludes certain people from access to assisted reproductive services they need to be considered and discussed and, ultimately, answered. This is so for both the welfare of the child model and for the presumption of access model.

Conclusion

I suggest that as a society we need to consider whether regulation of access to assisted reproductive services is appropriate to prevent harm to a child who may be born as a result. Notwithstanding the extensive inquiries into assisted reproductive services that have been held over the years, many of the issues have not been adequately grappled with. This has often been because it has been assumed that by limiting access to married or de facto heterosexual couples, we are protecting children from harm. Yet such an assumption is flawed - it is both over-inclusive and under-inclusive. This assumption is grounded in inaccurate stereotypes: that single women and lesbians will not make good parents and that heterosexual couples will make good parents. In fact there are good and bad lesbian and gay parents, just as there are good and bad heterosexual parents and good and bad single parents. We thus need to adopt an approach to regulation of assisted reproductive services that actually attempts to consider the individual circumstances of those seeking access - be they heterosexual couples, lesbian couples or single women. Such a system would accord with our international legal obligations to prevent discrimination on the basis of sex, sexual preference and marital status and to act to further the best interests of the child.

I have raised various questions about whether we should have regulation at all and, if so, the kind of regulation that might be appropriate. I suggest that a system based on a "welfare of the child" test is inappropriate, because it is vague and subjective. A system based on a presumption of access is preferable, with specified grounds of exclusion - and exclusion should be subject to review by an appropriate body and, ultimately, by the courts. However, even this method may result in excluding some persons who ought not to be excluded, and the grounds of exclusion are by no means unproblematic. Ultimately, I hope to spark debate on these issues rather than to offer a prescription for what must be done. But one thing is clear - any system of regulation based on discrimination needs to change.

Notes

1 *McBain v Victoria* (2000) 99 FCR 116. A more detailed analysis of the *McBain case* can be found in Walker, 2000.

2 That is, women not married or in a heterosexual de facto relationship. As a matter of law this includes lesbian women, although many lesbians are in relationships and thus not appropriately described as single.

3 As a matter of Australian constitutional law, under s 109 of the Constitution, where a State law is inconsistent with a Commonwealth law, the latter prevails and the former is inoperative to the extent of the inconsistency.
4 See *Pearce v South Australian Health Commission* (1996) 66 SASR 486.
5 Office of the Prime Minister, Media Release, *Amendment To Sex Discrimination Act*, 1 August 2000, http://www.pm.gov.au/news/media_releases/2000/disc_act1-8.htm.
6 *Re Sundberg; Ex parte Australian Catholic Bishops Conference*, High Court of Australia, Matter Nos C22/2000, C6/2001, heard 4-6 September 2001, decision pending.
7 See, eg, Commonwealth of Australia, Australian Health Technology Advisory Committee, *Review of Assisted Reproductive Technology* (1998).
8 See, eg, Victoria, The Committee to Consider the Social, Ethical and Legal Issues Arising from In Vitro Fertilisation, *Consolidated Reports of the Victorian Inquiry Into IVF and Related Issues* (1990).
9 See, eg, Queensland, *Report of the Special Committee Appointed by the Queensland Government to Enquire Into the Laws Relating to Artificial Insemination, In Vitro Fertilisation and Other Related Matters*, Vol 1, (1984); Tasmania, Committee to Investigate Artificial Conception and Related Matters, *Final Report* (1985).
10 There are exceptions that have considered the interests of the child in some greater detail. See: Commonwealth of Australia, National Bioethics Consultative Committee, *Access to Reproductive Technology* (1991); New South Wales, Law Reform Commission, *Report 49 – Artificial Conception: Human Artificial Insemination* (1986); Western Australia, Select Committee on The Human Reproductive Technology Act 1991, *Report* (1999).
11 See, eg, Misha Schubert, "Bracks Ban Misguided: IVF Leader" *The Australian*, 3 August 2000.
12 Current legislative regimes in Australia provide for extensive information and counseling to be provided to donors, to insure that the decision to donate is indeed an informed one. See, eg, Victoria: *Infertility Treatment Act*, ss 16, 17.
13 *JM v QFG and Ors*, Queensland Anti-Discrimination Tribunal, Reasons for Decision, RG Atkinson, 31 January 1997, p. 8 (copy on file with the author).
14 International Covenant on Civil and Political Rights, Art 26.
15 *Infertility Treatment Act*, ss 11, 16, 18.
16 Australian adoption legislation currently discriminates against lesbians and gay men and single people in that a single person may only adopt in "special circumstances" and a same-sex couple may not adopt as a couple at all: see, eg, *Adoption Act* 1984 (Vic), ss 10, 11.
17 This is made clear in the Human Fertilisation and Embryology Authority's (HFEA) *Code of Practice* (5th ed), Clause 3.3 (http://www.hfea.gov.uk/frame.htm).
18 United Kingdom, *Hansard*, House of Lords, 5 Dec 1996, col 823 (Baroness Warnock).
19 HFEA Code of Practice (5th ed), http://www.hfea.gov.uk/frame.htm, clause 3.13.
20 HFEA Code of Practice (5th ed), http://www.hfea.gov.uk/frame.htm, clause 3.14.
21 *Ibid*, clause 3.15.
22 Judicial review may already be possible in the UK, although this was not fully decided as the case was dismissed on other grounds: see *R v Ethical Committee of St Mary's Hospital (Manchester); Ex parte H* [1988] 1 FLR 512.
23 This list substantially reflects the access criteria used in South Australian under the Council for Reproductive Technology's Code of Ethical Practice, except that that Code also includes an exclusion based on conviction for any offence involving violence. The

Code has been embodied in regulations and thus has the force of law: *Reproductive Technology (Code of Ethical Clinical Practice) Regulations* 1995 (SA).

24 This is prohibited in Victoria and the UK, for example, and has recently been the subject of controversy in both jurisdictions. See Lee and Morgan (2001, p.131); South Australian Council on Reproductive Technology, *Media Summary*, http://www.dhs.sa.gov.au/reproductive_technology/media-summary.asp.

25 *R v Ethical Committee of St Mary's Hospital (Manchester); Ex parte H* [1988] 1 FLR 512.

26 See above n 23.

27 See, eg, South Australian Council on Reproductive Technology, *Annual Report* 2000, 12.

References

Acton, H.B. (1972) (ed), *John Stuart Mill: Utilitarianism, On Liberty and Considerations on Representative Government*, Everyman, London.

Blyth, E. & Cameron, C. (1998), 'The Welfare of the Child: An Emerging Issue in the Regulation of Assisted Conception' *Human Reproduction*, Vol. 13, p.2339.

Burghes, L. (1996), 'Debates on Disruption: What Happens to the Children of Lone Parents?' in Elizabeth Bortolaia Silva, *Good Enough Mothering? Feminist Perspectives on Lone Mothering*, Routledge, New York, p.157.

Daniels, K. & Taylor, K. (1993), 'Formulating Selection Policies for Assisted Reproduction' *Social Science & Medicine*, Vol. 37, p.1473.

Dworkin, R. (1987), 'What Is Equality? Part 3: The Place of Liberty' *Iowa Law Review*, Vol. 73, p.1.

Dworkin, R. (1993), *Life's Dominion*, Knopf, New York.

Golombok, S. (2000), *Parenting: What Really Counts*, Routledge, Philadelphia, PA.

Harris, J. (1998), 'Rights and Reproductive Choice' in John Harris and Soren Holm, *The Future of Human Reproduction*, Oxford University Press, New York, p.5.

Hart, H.L.A. (1968), *Law, Liberty And Morality*, Oxford University Press, Oxford.

Jansen, R. (1997), 'Evidence-Based Ethics and the Regulation of Reproduction' *Human Reproduction*, Vol. 12, p.2068.

Lee, R.& Morgan, D. (2001), *Human Fertilisation and Embryology: Regulating the Reproductive Revolution*, Blackstone, London.

Mnookin, R. (1975), 'Child Custody Adjudication: Judicial Functions in the Face of Indeterminacy' *Law and Contemporary Problems*, Vol. 39, p.226

Mumford, S.E., Corrigan E. and Hill M.G.R. (1998), 'Access to Assisted Conception: A Framework of Regulation' *Human Reproduction*, Vol. 13, p.2349.

O'Neill, M. (1994), 'Dangerous Families' in Freda Briggs (ed), *Children and Families: Australian Perspectives*, Allen & Unwin, St Leonards, NSW, p.126.

Parks, C. (1998), 'Lesbian Parenthood: A Review of the Literature' *American Journal of Orthopsychiatry*, Vol. 68, p.376.

Pennings, G. (1999), 'Measuring the Welfare of the Child: In Search of the Appropriate Evaluation Principle' *Human Reproduction*, Vol. 14, p.1146.

Robertson, J. (1994), *Children of Choice: Freedom and the New Reproductive Technologies*, Princeton University Press, Princeton.

Smart, C. (1989), *Feminism and the Power of Law*, Routledge, New York.

Stern, J., Cramer C., Garrod A., and Green R. (2001), 'Access to Assisted Reproductive Technology Clinics: A Survey of Policies and Practices' *American Journal of Obstetrics and Gynaecology*, p. 591.

Waldron, J. (1993), *Liberal Rights: Collected Papers 1981-1991*, Cambridge University Press, Cambridge.

Walker, K. (2000), '1950s Family Values vs Human Rights: IVF, Donor Insemination And Sexuality In Victoria' *Public Law Review*, Vol. 14, p.292.

Chapter 12

Lesbian Couples In DI Practice

Anne Brewaeys

Introduction

As a result of an increasing tolerance towards homosexuality during the past 20 years, a growing number of lesbian women have come out for their sexual orientation and consequently lesbian motherhood came into society's attention. Up until now the largest numbers of lesbian mothers have had children who were born in the context of a heterosexual relationship, with both a biological mother and father. The lesbian 'coming out' process of those mothers occurred after the birth of the children, often resulting in the divorce of the parents. Recently, however, more and more lesbian women have decided to become pregnant within a lesbian relationship. The vast majority has used donor sperm. The status of the donor himself varies among the families involved. When the couple applies to an official fertility centre, donor choices depend on the actual policy of the clinic. In the past, most fertility centres in Western Europe have used anonymous donors. It is only recently that, for example in the Netherlands, identity registered donors are available. Thus most of the children already born in these lesbian families will have no access to donor information. There is, however, in the lesbian community a clear tendency towards the use of a known donor, a friend or relative, very often gay himself. When self-insemination techniques are used an application to a fertility service will only occur when one of the candidate parents is confronted with a fertility problem. Adoption is another way for lesbian women to share their life with children but one has to take into account that only a minority of western countries accepts non-traditional couples in their adoption programs. Thus lesbian women find a variety of ways to share their life with children and only a part of them makes use of official fertility services. It is however this group that will be in focus in this paper.

Whether or not to give access to lesbian couples in fertility clinics is a question still under debate in several Western countries. Despite the increased acceptance of same sex relationships, reluctance remains with regard to lesbian motherhood. It has often been argued that a child needs a father for healthy psychological development, that lesbian mothers are less maternal and that their children would be at risk for disturbances in emotional, gender and social development. The acceptance of lesbian couples in Donor Insemination (DI) programs is strongly dependent on the cultural, religious and political attitudes towards homosexuality in a certain society. Guidelines and practices vary a great deal. For example, the English Human Fertilisation and Embryology Act discourages insemination of lesbian couples by the importance attached to the child's need for a father (Morgan & Lee, 1991). In the Netherlands on the other hand, there is now an official guideline from the Ministry of Health in which exclusion of lesbian couples from infertility services is regarded as an act of discrimination.

This papers reviews findings of the most relevant empirical research and, based on the conclusions of this research, a number of recommendations regarding the practice of Donor Insemination with lesbian couples are formulated.

Findings of empirical studies

The controversies surrounding the subject of lesbian motherhood has stimulated researchers to investigate the consequences of being raised in a lesbian family. Since the need for evidence based clinical practice is widely accepted these days, the findings of this empirical research are of relevance to the actual debate. An overview will be presented on the following topics: characteristics of lesbian women applying to fertility centres, lesbian mother families with young children and children's own experience on being raised in a lesbian mother family.

Characteristics of lesbian women applying to fertility centres

Several studies have gathered information on the women's demographic features, their donor choices, their concept of future parental roles and their lesbian identity. Results revealed a very similar picture of lesbian women applying at fertility services in Europe and North America (Englert, 1994; Brewaeys et al., 1995; Baetens et al., 1996; Leiblum et al., 1995; Gartrell et al., 1995; Wendland et al., 1996; Jacob et al., 1999).

The majority was highly educated and white, had stable and long lasting relationships, lived openly as a lesbian and was accepted as such by significant others. These findings show that lesbian women visiting fertility centres had relatively privileged social and economic positions. Findings with regard to their parental skills may therefore not be extrapolated to the total population of lesbian couples.

When compared with heterosexual DI applicants, a significant greater proportion of lesbians was opting for an identity registered donor, if this choice was available. Obviously this choice was inspired by the fact that all lesbian mothers intend to reveal the use of a donor to their children in an early developmental stage. Future parental roles were intended to be equally divided between the two women. They considered the future child as 'their' child and described their family as a two mother unit.

Lesbian mother families with young children

In the past twenty years a number of studies have been carried out in lesbian families where the mother had lived in a heterosexual marriage before adopting a lesbian identity. Several child and family characteristics were investigated in these post divorce lesbian families and no adverse effects of the mothers' sexual orientation could be identified (for review see Brewaeys & van Hall, 1997).

More recently, investigations have been carried out among young children who have been conceived by means of DI within a lesbian relationship. These newly created families differ from the post divorce lesbian households on a number of important characteristics. There has never been a father figure and in most cases an anonymous donor was used. The children had been raised from birth by two lesbian mothers. Since the psychological development of young children is generally believed to be influenced by the home environment, follow up of these families is highly relevant.

The studies mentioned here are restricted to those in which a comparison group of heterosexual parents was used (Steckel, 1987; Flaks et al., 1995; Brewaeys et al., 1997; Golombok et al., 1997; Chan et al., 1998a, 1998b). The results with regard to family relationships and the psychological development of children appeared to be strikingly unanimous.

Family relationships Interesting to know is how these lesbian mothers have brought their new parental roles into practice. Most looked at was the division of educational tasks between the two mothers, attachment relationships and the quality of the mother-child interactions.

Child care and professional activities were more equally divided between the lesbian mothers than between heterosexual mothers and fathers. Involvement in educational tasks was higher for the social mothers than for the fathers. The attachment behaviour between the child and the social mother was similar to that between the child and the heterosexual father. One study even found that children of lesbian families were more securely attached than children of heterosexual parents (Golombok et al., 1997). Overall, lesbian mothers showed a higher quality of parent-child interaction compared with the heterosexual two headed families. Interestingly, the social mothers were doing better than the heterosexual fathers involved. However, the authors unanimously conclude that these parental strengths of lesbian mothers have more to do with their gender than with their homosexuality as such.

Psychological development of children As it is generally believed that the home environment influences the well being of children, the following aspects of child development have been investigated: gender development, emotional/behavioural problems and self esteem. No significant differences could be found between children raised in lesbian mother families and children raised in heterosexual families for the variables mentioned above. Thus, there is no empirical evidence for a possible negative or positive effect of lesbian motherhood on child development.

Children's experience on being raised in a lesbian mother family

So far, the overall picture of lesbian mother families is a positive one. These children have been growing up for the first years of their lives in a warm and secure family environment and consequently, children were doing well for all variables studied. However, because of the young age of the children involved, these studies do not tell us what it means for the children themselves to grow up in a lesbian family. There are at least two important questions that need further investigation: what does it mean for children to be the offspring (1) of a lesbian mother and (2) of a sperm donor?

Once children reach school age, they begin to move away from their parents and contact with peers becomes more important. Young adolescents in particular may pressurise each other to conform to the values of the peer group. Undoubtedly these children will have to cope actively with homophobic attitudes. The question is then whether they will still be as accepting towards their non-traditional family structure. Furthermore, a young child cannot fully understand the consequences of being the offspring of a donor, in most cases an anonymous one. Children's reactions to the knowledge that they are genetically linked to a stranger are still unknown since their donor concept has not yet been investigated.

Findings about adolescent and adult offspring raised by lesbian mothers remain extremely scarce. There are two studies, however, which may give us some insight into the questions mentioned above since they report on the experiences of the children themselves.

Being the offspring of a lesbian mother The longitudinal study of Tasker and Golombok (1997) reports findings on adults who grew up in a lesbian family. As all children were born into a heterosexual relationship before the lesbian coming out of their mother, they all knew who their father was. Twenty five young adults (mean age 23.5 years) were interviewed about several aspects of their family life. Their experiences were compared with those of adults raised in a post divorce heterosexual single mother family. Data were obtained on their family relationships, their peer relationships, their histories of intimate relationships and their psychological adjustment.

Adult children of lesbian mothers were generally more positive about their *family relationships* than were those brought up in heterosexual post divorce families. Positive feelings about living in a lesbian family were associated with a more open lesbian life style of the mother, and with a higher quality of the mother child relationship. When the mother had a partner, the children of lesbian mothers gave more positive accounts of their relationship with their mother's female partner than did those in the control group in relation to their stepfather. However, during adolescence, feelings of embarrassment about having a lesbian mother were frequently reported. Many interviewees remembered that they disliked their mothers being too open about their sexual orientation in the presence of their friends.

With regard to their *peer relationships*, children raised in lesbian mother families did not recall more peer stigma during adolescence than children in heterosexual families. Furthermore, they were no more likely to recollect being teased about their mother. However, young people from lesbian households remembered more often peer group teasing about their own sexual identity and this was especially so for the boys. Many also reported having felt some concern that if their peers found out about their family background, they might be hostile. Of them 66 per cent had not told their friends that their mother was a lesbian.

Since one of the often heard assumptions with regard to lesbian motherhood is that their children would have a greater chance to become lesbian or gay themselves, these young adults' histories of *intimate relationships* were registered. The quality of intimate relationships, the number and duration of such relationships did not differ between children of lesbian and heterosexual single mothers. Moreover the number of people identifying themselves as gay or lesbian was similar in both groups. Children of lesbian mothers, however, reported more same sex experiences than children of heterosexual mothers. Furthermore, lesbian mothers were more able to communicate about sexual matters with their children than heterosexual mothers. Although the authors did not rule out the influence of biological factors, they argued that it was the open family attitude about same sex relationships that primarily influenced the young person's same gender sexual interest.

The *psychological adjustment* of these young adults appeared to be healthy. Men and women raised by lesbian mothers were no more likely than the control group to experience psychological problems. Depression and anxiety scores of both groups fell within the normal range.

In summary, adults who grew up in a lesbian family were positive about their non-traditional family identity, they appeared to be well adjusted and did not recall peer teasing with regard to their family structure. There was however a clear tendency, especially during adolescence, to be secretive about their mothers' homosexuality. This must undoubtedly have been related to the perceived homophobia within their social network.

Being the offspring of an anonymous donor The data presented here are part of a longitudinal study carried of lesbian families who conceived after having used an anonymous donor. The actual survey was carried out among 44 children with a mean age of 10 years (range 8 to 18 years). For the first time the children themselves participated actively in the study and could tell us what it meant to grow up in a lesbian DI family (Vanfraussen et al., 2001).

In contrast to heterosexual DI families, these lesbian mothers had informed their children in an early developmental stage. Children were interviewed about their birth concept when they were 5 years old and they were well aware that their two mothers had used donor seeds in order to conceive and that there was no father within the family (Brewaeys et al. 1997). However, the question remained what these children would think of their donor origin once they had reached a more advanced stage of cognitive development. We assumed that after 8 years of age, children were capable of making the link between 'the use of the seeds' and the man behind the seeds to whom they were genetically related. We first registered what children would have liked to know about the donor and also interviewed their mothers about this issue. Then we investigated if the need to know more about the donor was associated with certain child or family characteristics.

With regard to their *donor choices*, 41 per cent of the boys and 10 per cent of the girls said that, if they had the occasion, they would have liked to meet the donor. The great majority of the children would have liked to meet the donor in order to check whether they found any resemblance between themselves and the donor. Two children however, were in search for a father figure.

Nine per cent of the boys and 32 per cent of the girls were happy with non-identifying information. Most frequently they wanted to know if the donor looked like them, or what kind of personality he had. The remaining children did not need any information at all.

Our data seem to support the assumption that, if children are aware of their donor status, at least some of them will want to meet the donor. Whether or not similar results would have been found in a representative heterosexual DI sample remains an open question. Would the presence of a social father have influenced the children's responses? An intriguing question in this regard is whether the boys would still outnumber the girls in their wish to meet the donor. The fact that there is no father figure to identify with in the lesbian families, might have caused the gender difference found in this study. Moreover, one has to take into account that the children interviewed in this study had a mean age of 10 years. Adoption literature points out that questions about one' s biological parents increase with the age of the adopted child. Is it therefore reasonable to expect the same trend in our study group? A further follow up study will have to provide us with the answer.

The need to know more about the donor did not appear to be associated with *certain family or psychological characteristics*. A comparison between children who wanted to know more and children who did not, failed to reveal significant differences with regard to the quality of the mother-child interactions, children's self esteem and behavioural/ emotional problems. The lack of such an association contrasts with some findings of the adoption research where the need to know more about one's biological parents is related with less psychological well-being and poorer family relationships.

With regard to their donor choices, *mothers* held different views. Compared with their children, they more often reported to want no information at all about the donor (79 vs. 54 per cent). Paradoxically enough, it seems that, at the stage that a number of children start to show some interest in the donor, mothers are experiencing him as a threat to their family unit and want to disregard him as a person.

Of course, one has to take into account that the knowledge of having used an anonymous donor, at the time an irreversible choice, might have influenced mothers' attitudes as well. Nevertheless, the question remains whether such a discrepancy between the interests of children and mothers may lead to communication problems within the family. One could also expect that children, in their attempt to be loyal towards their mothers, might get isolated with their wish to know more about the donor. Moreover, taking into account that none of these children will ever be able to gather whatever donor information, a potential risky situation is created.

Conclusion

Having reviewed these follow up studies, what are the most salient features of lesbian mothers and their children? And, based on these findings, should lesbian couples get access to fertility services?

Lesbian mothers with young children showed a high quality of parenting and the social mothers had even more parental skills than the fathers in the heterosexual families. Their children were equally attached to both of them and appeared to be psychologically well adjusted. In sharp contrast with this picture is the overall view of society, in which a lesbian household is still regarded as a single mother family. Important to note in this regard is the vulnerability of the social mothers' parental role. Although accepted by her family members as a parent, her position outside the family does not exist. There are no rules, no symbols, not even a name to make her visible in society. The best solution to this problem is to legalise her position. This is what happened last year in the Netherlands. As the only country in the world, the official institute 'marriage' is open for same sex couples: they have now the same rights and duties as every other heterosexual married couple. Moreover, the legal possibilities are available to formally become equal parents with identical rights and duties towards the children.

Undoubtedly, dealing with being a non traditional family may induce stress to all family members. Little is known about the impact of homophobic reactions on the well-being of mothers and children. The few available findings on the children's own experiences reveal that adolescents are sometimes secretive about their mothers' sexual orientation. However, findings of the same study also seem to suggest that children are best off with a mother who is open about her sexual orientation, disposes of an accepting social network and possesses good parental skills. Fertility workers should take this into account when counselling lesbian couples.

When children of lesbian mothers were asked what they would have liked to know about the donor, one quarter of them answered that they would have liked to meet him. Having informed their children in an early developmental stage enhanced their offspring's wish to know more about the donor.

Accepting lesbian couples has therefore important implications for DI practice. Medical doctors are now confronted with women applying for DI, not because of the infertility of their male partner but because of the absence of such a partner. Obviously, when treating lesbian couples, it can no longer be denied that DI is not a medical treatment but an alternative route to parenthood. Further research is needed in order to investigate these children's donor concept. Meanwhile, it is clear that DI programs with lesbian couples should only use identifiable donors. Adequate psychosocial counselling programs need to be developed in order to help mothers and children in their possible search for donor information.

And finally, when looking at these empirical findings, there seem to be no good reasons to exclude lesbian couples from DI programs. Of course, results remain preliminary and most of the families investigated were white, economically privileged and surrounded by a relatively tolerant social network. Many questions remain unanswered and large scale longitudinal studies are still needed. But on the other hand, what do we know about the impact of DI on children born in heterosexual families? The results of the few studies carried out so far revealed less unanimous findings than those conducted in lesbian DI families (for review see Brewaeys, 2001). Nevertheless, thousands of these children were born all over the world. The reason for this divergent approach is simple, heterosexuality is the norm, homosexuality is deviant from the norm. So, the answer to the question as to whether lesbians should get access to DI greatly depends on one's view with regard to homosexuality. Ethnic, cultural and religious differences influence the degree of tolerance towards gays and lesbians in a certain society. The importance of this broader context in which fertility services operate, underlines that there exists no universal answer.

References

Baetens, P., Ponjaert-Kristoffersen I., Van Steirteghem A.C. and Devroey P. (1996) 'Counselling lesbian couples in creating a family by artificial insemination by donor', *Human Reproduction*, Vol. 11 (abs.), p. 79.

Brewaeys A. (2001), 'Parent-child relationships and child development in donor insemination families', *Human Reproduction Update*, Vol. 7, pp. 38-46.

Brewaeys A., Ponjaert I., van Hall E., Helmerhorst F. and Devroey P. (1995) 'Lesbian mothers who conceived after DI, a follow up study', *Human Reproduction*, Vol. 10, pp. 2731-35.

Brewaeys A., Ponjaert I., van Hall E. and Golombok S. (1997), 'Donor Insemination: Child and family development in lesbian-mother families with children of 4-to 8 years old', *Human Reproduction*, Vol. 12, pp. 1349-1359.

Brewaeys A. and van Hall E. (1997), 'Review: Lesbian mother families: The impact on Child and Family Development', *J. Psychosom. Obstet. Gynecol.*, Vol. 18, pp. 1-17.

Chan, R., Raboy B. and Patterson C.J. (1998), 'Psychosocial adjustment among children conceived via donor insemination by lesbian and heterosexual mothers', *Child Development*, Vol. 69, pp. 443-457.

Chan, R., Risa C., Raboy, B. and Patterson, C.J. (1998), 'Division of labor among lesbian and heterosexual parents: associations with children's adjustment', *Journal of family Pychology*, Vol. 12, pp. 402-419.

Englert, Y. (1994), 'Artificial insemination of single and lesbian women with donor semen', *Human Reproduction*, Vol. 9, pp. 1969-1971.

Flaks D., Fisher I., Masterpasqua F. and Joseph G. (1995), 'Lesbians choosing motherhood: a comparative study of lesbian and heterosexual parents and their children', *Developmental Psychology*, Vol. 31, pp. 105-114.

Gartrell N., Hamilton J., Banks A., Mosbacher D., Reed N., Sparks C. and Bishop H. (1995), 'The national lesbian family study: 1. Interviews with prospective mothers', *American Journal of Orthopsychiatry*, Vol. 66, pp. 272-281.

Golombok, S., Tasker F. and Murray C. (1997) 'Children raised in fatherless families from infancy: family relationships and the socio-emotional development of children of lesbian and single heterosexual mothers', *J. Child Psychol. Psychiatry*, Vol. 38, pp. 783-791.

Jacob M. (1999), 'Lesbian couples and single women', in Burns, L. and Covington, S. (eds) *Infertility counselling. A comprehensive handbook for clinicians*, Parthenon, New York.

Leiblum S., Palmer M. and Spector I. (1995), 'Non traditional mothers: single heterosexual/lesbian women and lesbian couples electing motherhood via donor insemination', *J. Psychosom. Obstet. Gynecol.*, Vol. 15, pp. 11-20.

Morgan, D. and Lee, R.G. (1991), *Blackstone's Guide to the Human Fertilisation and Embryology Act*, Blackstone Press Limited, London.

Steckel A. (1987), 'Psychological development of children of lesbian mothers', in Bozett F.W. (ed) *Gay and lesbian parents*, Praeger New York, pp. 75-87.

Tasker F. and Golombok S. (1997), *Growing up in a lesbian family*, Guilford Press, New York.

Vanfraussen K., Ponjaert I. and Brewaeys A. (2001), 'An attempt to reconstruct children's donor concept: a comparison between children's and lesbian parents' attitudes towards donor anonymity', *Human Reproduction*, Vol. 16, pp. 2019-2025.

Wendland, C., Byrn F. and Hill C. (1996), 'Donor insemination: a comparison of lesbian couples, heterosexual couples and single women', *Fertil. Steril.*, Vol. 65, pp. 764-770.

Chapter 13

Fertility Rights: Will The UK's Human Rights Act Make Any Difference To Access To Assisted Conception Treatment?

Eric Blyth

Introduction

Access to assisted conception treatment in the UK is governed by Section 13 (5) of the Human Fertilisation and Embryology Act 1990 (hereafter 'the Act') under which treatment must not be provided for a woman 'unless account has been taken of the welfare of any child who may be born as a result of the treatment (including the need of that child for a father), and of any other child who may be affected by the birth'. The Act does not define the welfare of the child although, through its *Code of Practice*, the regulatory body, the Human Fertilisation and Embryology Authority, has elaborated on the concept of welfare, emphasizing the parenting capacity of people seeking treatment (Human Fertilisation and Embryology Authority, 2001a). The Act does not afford any priority to the welfare of the child to be born in relation to the welfare of any other child or that of people seeking treatment. Neither does the Act specify what action - if any - should be taken after account has been taken of the child's welfare.

The origins of the Act's welfare provisions lie in the efforts of some legislators to prevent lesbians and single women from accessing assisted conception treatment and a desire to establish the *paramountcy* of a welfare requirement. In evidence to the Warnock Committee, a Committee of Inquiry established by the government to investigate human fertilization and embryology services and whose report provided the basis for the Act (Department of Health and Social Security (1984), the All Party Lords and Commons Family and Child Protection Group (1983) claimed that 'the interests and well being of any possible child must be the *principal* consideration' (my emphasis) and referred to 'the undesirability from the child's point of view of making AID ["artificial insemination by donor", the term by which donor insemination was then known] available to single women, whether unmarried, widowed or divorced, and lesbians'. While the Warnock Committee chose not to incorporate the Group's views in its own recommendations and, indeed, the Human Fertilisation and Embryology Bill that was laid before Parliament in 1989 contained no specific reference to the welfare of the child, the legislators had the opportunity to raise their concerns again during parliamentary debate on the Bill. At that time a Conservative backbench Member of Parliament, Jill Knight, was both Chairman of the Group and a member of the House of Commons Standing Committee that considered in detail the provisions contained in Bill. It was during Committee stage that the substance of what became S 13 (5) of the Act first emerged and its eventual composition represented a compromise that reflected both the parliamentarians' wishes and the government's existing obligations. The government observed that making the welfare of the 'child' paramount (the status it is afforded in UK child welfare legislation) - and thereby subordinating the welfare of others - risked contravening the European Convention for the Protection of Human Rights and Fundamental Freedoms - to which the UK was a signatory (Council of Europe, 1950; Bottomley, 1990a). Similar considerations lay behind the government's unwillingness to impose a general exclusion on single women and lesbians (Bottomley, 1990b) - despite the government's noted distaste for alternative, non- conventional family structures. The welfare requirement's otherwise anomalous reference to a child's need for a father is explicable, therefore, as political expedience - and UK legislators never anticipated that the child's need for a mother might become problematic.

In comparing the position in the UK with that of other jurisdictions where limitations may be placed on access to assisted conception treatment, it is also important to note that the Act does not require people seeking treatment to meet any requirements regarding their marital status or sexual orientation. Neither is access to assisted conception treatment dependent on evidence of impaired reproductive capacity or a diagnosis of clinical infertility.[1]

While the Act does not exclude 'any category of woman from being considered for treatment' (Human Fertilisation and Embryology Authority, 2001a, para. 3.10, p. 16), the HFEA requires that the need to take account of the welfare of the child will be 'greater if the treatment is required to be licensed under the HFE Act and particularly if it involves the use of donated gametes' (Human Fertilisation and Embryology Authority, 2001a, para. 3.9, p. 16). This requirement exerts a greater impact on single women and lesbians than heterosexual couples since treatment with donated gametes - and in particular donor insemination - is the form of treatment they are most likely to seek and available evidence indicates that discrimination in treatment centers against single women and lesbians is prevalent (Douglas, 1992, 1993; Englert, 1994; Saffron, 2001; Deech, undated). More recently, concerns have been expressed that assisted conception treatment centers are discriminating against HIV positive men and women (Apoola *et al.*, 2001).

Further restrictions on access to assisted conception treatment are governed neither by the Act itself nor the policies of the HFEA, but through the administration of publicly funded assisted conception treatment. Public funding of assisted conception treatment is generally limited and, although designated as a *National* Health Service, the UK's NHS is, in reality, composed of regional funding bodies. Each of these has responsibility for a designated geographical area, implementing its own priorities for health service provision and resulting in differentiation in the availability of publicly funded assisted conception services. So, for example, different limits may be placed on the type of treatment and the number of treatment cycles that may be publicly funded. The length of waiting lists for treatment may also vary. Local funding authorities may establish their own criteria – independently of the requirements of the Human Fertilisation and Embryology Act, since they are not bound by its provisions - for determining eligibility to treatment, e.g.:

1. the couple must be in a durable heterosexual relationship of at least three years' duration;
2. the couple must have no existing child - including an adopted child;
3. neither partner must previously have been sterilized;
4. the final treatment cycle must commence before the female partner reaches the age of 37 (North West Lancashire Health Authority, undated).

Some NHS funding authorities may impose even more stringent criteria. For example, some apply strict eligibility criteria relating to the age, sexual orientation and marital status of applicants and may exclude from publicly-funded treatment couples where *either* partner has a child from a previous relationship - and some may decide not to fund assisted conception treatment at all (College of Health, 1993; Wiles and Gordon, 1994; Wiles and Patel, 1995; Wiles and Oddos, 1996; Stone and Reisel, 1997; Kennelly and Reisel, 1998). In response to the current state of publicly-funded assisted conception treatment, Clare Brown, Executive Director of 'Child' -

one of the UK's national infertility patient support groups - and President of the International Federation of Infertility Patient Associations, has claimed that '...subfertility treatment on the NHS is not just rationed: it is patently arbitrary and totally unethical' (Brown, 2000, p. 268).

In addition treatment centres appear to apply different eligibility criteria depending on the source of treatment funding. Centres providing NHS treatment are more likely than their private counterparts to operate a rationing policy and impose more restrictive eligibility criteria. In contrast, private centres may provide treatment to whomsoever they choose with virtual impunity (Brazier, 1999; Bennett and Harris, 2001; Levitt, 2001; Rogers, 2001).

The European Convention for the Protection of Human Rights and Fundamental Freedoms and access to assisted conception treatment

The European Convention for the Protection of Human Rights and Fundamental Freedoms is a Treaty of the Council of Europe that came into force in 1953. In principle, UK citizens have enjoyed the rights guaranteed under the Convention since 1966 and the Convention clearly had some impact on the framing of the Human Fertilisation and Embryology Act as indicated above. However the requirement to seek a remedy to an alleged breach of a right guaranteed under the Convention through the Council of Europe's administrative and legal machinery meant that, in practice, seeking a remedy was both expensive and time consuming and was, therefore, infrequently undertaken by individual citizens. In October 2000, the Human Rights Act 1998, which had the effect of making rights guaranteed under the European Convention available in domestic courts, became law throughout the UK. The Act generally requires a public authority to ensure that its actions are compatible with the Convention. A key effect of the Human Rights Act has been to increase the accessibility of the Convention rights to UK citizens. Articles 8, 12 and 14 of the Convention are particularly relevant to assisted conception treatment.

Article 8 - The Right to Respect for Private and Family Life - provides that:

> 1. Everyone has the right to respect for his private and family life, his home and his correspondence.
> 2. There shall be no interference by a public authority with the exercise of this right except such as is in accordance with the law and is necessary in a democratic society in the interests of national security, public safety or the economic well-being of the country, for the prevention of disorder or crime, for

> the protection of health or morals, or for the protection of the rights and freedoms of others.

Article 8 is a qualified rather than an absolute right, and may be limited by virtue of the 'legitimate aims' that are outlined in paragraph 2. In deciding whether an alleged breach of a qualified right has occurred, two key principles must be taken into account by the court. First, the 'margin of appreciation' allows a member state a measure of flexibility and freedom in determining public policy and affords respect and sensitivity to member states' political and cultural traditions. A court will therefore, limit its own competence to judge the issue, so long as a permitted interference with the right is lawful, serves a legitimate purpose, is necessary in a democratic society, and is not discriminatory. The second principle, 'proportionality', serves to limit the degree of interference in pursuit of a legitimate aim to that which is no greater than is absolutely essential to meet a necessary policy objective:

> Inherent in the whole of the Convention is a search for a fair balance between the demands of the general interest of the community and the requirements of the protection of the individual's fundamental rights (*Soering v United Kingdom*, 1989).

Article 12 - Right to Marry and Found a Family - provides that: 'men and women of marriageable age have the right to marry and found a family, according to the national laws governing the exercise of this right'. Although Article 12 is not a qualified right in the same way that may be said of Article 8, since it contains no 'exception' clause, exercise of the right is dependent on relevant national laws, which typically prescribe the age at which a person attains 'marriageable age' and other criteria regarding eligibility to marry a particular individual, such as the prohibition of marriage between specified genetic relatives or the prohibition of same sex marriages. The concept of 'family' in Article 12 is narrower in scope than under Article 8 and relates to procreation within the conventional family unit (Rees v UK, 1987):

> The right to marry guaranteed by article 12 refers to the traditional marriage between persons of the opposite biological sex. This appears also from the wording of the article which makes it clear that Article 12 is mainly concerned to protect marriage as the basis of the family.

In addition to rulings of the Court itself, the view of the Council of Europe Steering Committee for Human Rights affords little comfort to unmarried heterosexual or gay couples or individuals looking to Article 12 as a means of advancing their rights to marry or to establish a family:

> the concern to protect the interest of the child ... can justify refusal of access to artificial procreation to persons not constituting a heterosexual couple or not living as a couple ... Any right to give life which might flow from Article 12 could only operate in the context of a couple consisting of a man and a woman (Council of Europe, 1989, p. 10).

Article 14 - The Prohibition of Discrimination - provides that: 'the enjoyment of rights and freedom set forth in this Convention shall be secured without discrimination on any grounds such as sex, race, colour, language, religion, political or other opinion, national or social origin, association with a national minority, property, birth or other status'.

Article 14 may only be invoked in conjunction with another provision of the Convention that establishes a right or freedom, although no breach of that other Convention right need be established. Once it is established that persons in similar situations have received different treatment, such a difference is discriminatory without an objective and reasonable justification.

Two cases involving alleged breaches of human rights guaranteed under the Convention in respect of assisted conception have already been determined at the European level. In the first a Dutch sperm donor alleged breach of Article 8 (J.R.M. v Netherlands, 1993). The second case (X, Y and Z v UK, 1995) involved a British female-to-male transsexual, Dr Stephen Whittle, who had undergone gender reassignment surgery and who alleged breach of Article 8 on the grounds that the UK government would not permit him to be registered as the father of four children born to his partner following donor insemination, to which he had consented under provisions of the Human Fertilisation and Embryology Act 1990.

In J.R.M. v Netherlands (1993) the applicant, a Dutch citizen, and resident in the Netherlands, had agreed to donate sperm to enable a lesbian couple, who wished to use a known donor because they considered it important for the child to know his or her father, to conceive a child. Following successful insemination a girl was born in July 1987 and the applicant regularly visited the child and the two women, and baby-sat on a weekly basis. In 1988 the applicant sought to establish certain visiting arrangements, i.e. that the child would spend one weekend per month with him, citing an alleged previous agreement between himself and the two women. The latter denied that any such agreement had previously been made and refused further contact between the applicant and the child. The applicant sought access to the child via the Dutch Juvenile Court

(Kinderrechter). The court declared that the applicant's role as a donor in itself was an insufficient basis for claiming the establishment of 'family life' within the meaning of Article 8 of the Convention, but that the applicant's request was admissible on the basis of the combination of the applicant's role as a donor and his weekly contact with the child over an eight month period. However, the court concluded that, given the conflict between the applicant and the two women, the applicant's access would be contrary to the child's interests. On appeal by the two women, the Dutch Court of Appeal (Gerechtshof) quashed both decisions of the Juvenile Court and declared inadmissible the applicant's request to have access to the child, taking the view that, even combined with his role as the child's genetic father, his actual contact with the child was of such a 'summary character' that it could not constitute a family life with the child within the meaning of Article 8 of the Convention. The applicant's subsequent appeal to the Dutch Supreme Court (Hoge Raad), and his argument that there is always 'family life' between a biological father, including a father by virtue of sperm donation, and his child, was rejected. When the European Commission of Human Rights considered the applicant's case in 1993 it, too, determined that his role as donor and his limited contact with the child were insufficient to demonstrate the existence of family life between him and the child, citing previous rulings that family life implies close personal ties in addition to parenthood and that the existence or non-existence of family life will depend on a number of factors, of which cohabitation is only one, and on the circumstances of each particular case. The applicant also claimed that the decision of the Dutch courts to declare his request inadmissible discriminated against him in comparison with a father of a legitimate child and that his rights under Article 14 of the Convention had been breached. The Commission's rejection of this claim was brief and to the point (if not particularly illuminating):

> In view of the fundamental differences between the applicant and a father of a legitimate child, the Commission does not find that these two situations can be compared and considered as being analogous and, therefore, no question of discrimination arises in the present case.

The facts of (X, Y and Z v UK, 1995) are as follows. At birth Dr Stephen Whittle had been registered as female and had undergone gender reassignment surgery. He had been living in a stable relationship for a number of years with a female partner who had given birth to two children born following donor insemination. Dr Whittle had consented to his partner's insemination under provisions of the Human Fertilisation and Embryology Act 1990. UK law defines a person's gender by reference to biological criteria at birth and does not recognise that it can be changed by gender reassignment surgery. As a result of this principle, a female-to-male transsexual is not permitted to marry a woman and cannot be acknowledged as the father of a child. Dr Whittle's claim that this policy violated his rights under Article 8 was declared admissible by the European Commission on Human Rights, which referred the case to the Court of Human Rights for final determination. In a majority 14-6 decision the Court itself determined against Dr Whittle since it took the view that there was no common European standard with regard to the granting of parental rights to transsexuals or how the social relationship between a donor conceived child and the person performing the social role of father should be reflected in law. Accordingly the Court concluded that the UK government should be afforded a wide margin of appreciation in interpreting the law. Dr Whittle's claim that his rights under Article 14 had also been breached since, had he been a legally-recognized male, under the provisions of the Human Fertilisation and Embryology Act 1990 he would have been permitted to register himself as the father of the children, was also rejected by the Court by a majority decision.

The Human Rights Act 1998

In addition to making rights guaranteed under the European Convention available in domestic courts, the Human Rights Act requires a public authority to ensure that its actions are compatible with the Convention, thus supplementing existing provisions for an individual wishing to complain that a public authority had acted unlawfully in carrying out its public functions through application for judicial review. In general, the grounds for judicial review are illegality, procedural impropriety and irrationality. The Human Rights Act provides an additional ground for court scrutiny of public authority decision-making, where a decision is alleged to be incompatible with a Convention right.

The Act's definition of a 'public authority' (Section 6(3)(b)) unambiguously includes Health Authorities, National Health Service Trusts and the HFEA. The status of a private assisted conception center is less clear, although there is a strong argument for defining a private clinic as a 'public authority' when providing treatment to NHS patients and it may also be covered when treating private patients, given that it operates in the public domain and exercises powers and duties of a public nature under a statutory scheme (Miles, 2001).

In anticipation of the introduction of the Human Rights Act and its potential impact on assisted conception treatment, Ruth Deech, Chair of the HFEA, is quoted in a national newspaper as follows:

> I suspect that from October we're going to have a lot of interesting questions being raised about whether treatment can be refused to single women, to women of 60, to lesbian and gay couples, to widows, and whether sex selection for social reasons can be denied ... I'm not saying that all the careful protection we have will crumble, but it seems to be likely that some very deep questions will be posed to the courts (cited in Boseley *et al.*, 2000).

In the same newspaper article, Baroness Warnock, who chaired the original Committee of Inquiry, is said to have described the law as 'a real can of worms', adding that:

> It never seemed to me the right to a family enshrined in European law had any reference to single people, post-menopausal women or homosexual couples ... It seems to me to be very risky to interpret the law in that way (cited in Boseley *et al.*, 2000).

In practice, within the first year of implementation, few cases concerning assisted conception have been brought under the Human Rights Act and, to date, none have troubled existing arrangements regarding access to treatment. (One case, ongoing at the time of writing, but which is outwith the scope of the present chapter, concerns a woman conceived following anonymous sperm donation, Jo Rose, and an unnamed five year old donor-conceived child, who have claimed that the restriction on information available about the donor breaches their rights under Article 8 [Peek, 2001]).

In 1999, the only daughter of Alan and Louise Masterton, died in a bonfire accident at the age of three. The Mastertons, who have four boys, wished to have another daughter and wanted to use pre-implantation genetic diagnosis (PGD) to choose the gender of their baby. However, the HFEA does not permit PGD for social, physical or psychological reasons. The Mastertons were reported to be considering legal action alleging infringement of their rights under Articles 6 and 8 (*BioNews*, 2000) although, in the event they were unable to find a UK clinic that was prepared to apply to the HFEA for a license variation to allow them to select the gender of their embryos. Consequently neither the HFEA nor the UK courts were ever asked to make a decision in this matter although Mr Masterton did complain about the HFEA to the parliamentary ombudsman on a point of procedure. The ombudsman found that whilst the HFEA had not acted unreasonably, Mr Masterton should receive an apology for having his expectations raised unnecessarily about the way in which the HFEA would process his request (HFEA, personal communication). In the meantime, debate about the use of PGD for embryo sex selection continues in the UK (e.g. *Bionews*, 2001).

In another case, a 46 year-old patient of Mohammed Taranissi, medical director of the Assisted Reproduction and Gynaecology Centre, London, and supported in her action by Mr Taranissi, challenged the HFEA-imposed limit of implanting no more than three embryos *in utero* in a single treatment cycle,[2] on the grounds that this would limit her chances of conceiving, and breached her rights under Articles 8 and 12. The High Court rejected this case in April 2001 a decision that was upheld on appeal by the Court of Appeal in January 2002 (R v Human Fertilisation and Embryology Authority *ex parte* (1) Assisted Reproduction and Gynaecology Centre (2) H, 2002).

The final case involved a man, Gavin Mellor, serving a life prison sentence for murder. By virtue of his incarceration and in accordance with prison regulations, he was prevented from engaging in sexual intercourse with his wife. Mellor's request that his wife be allowed to conceive by insemination with his sperm was rejected by the Home Office and Mellor claimed that this refusal breached his rights under Articles 8 and 12 (R v the Secretary of State for the Home Department *ex parte* Mellor, 2001). Citing an earlier ruling by the European Commission on Human Rights, the Court of Appeal noted that:

> Article 8 … does not mean that a person must at all times be given the actual possibility to procreate his descendants. It would seem that the situation of a lawfully convicted person detained in prison in which the applicant finds himself falls under his own responsibility, and that his right to found a family has not otherwise been infringed.

Commenting on the Mellor case specifically, Lord Phillips, MR, observed:

> It is not obvious that the signatories to the Convention would have agreed that a man who had, by imprisonment, been justifiably deprived of the enjoyment of family life and the exercise of conjugal rights, should be entitled to inseminate his wife artificially in order to produce a child in whose development and support he could play no part.

Lord Phillips expressed his concern about the deliberate creation of single parent families by means of assisted conception that might give an indication how courts might deal with any future case brought by a single woman or indeed, by women in a same-sex relationship:

> By imprisoning the husband the State creates a situation where, if the wife is to have a child, that child will, until the husband's release, be brought up in a single parent family. I consider it legitimate and indeed desirable, that the State should consider the implications of children being brought up in those circumstances when deciding whether or not to have a general policy of facilitating the artificial insemination of the wives of prisoners or of wives who are themselves prisoners.

Conclusion

On past and recent performance, those charged with interpreting the Convention rights have tended to err on the side of caution. Against that, the Convention is said to be a 'dynamic document', to be interpreted in the light of modern day circumstances. It remains to be seen how dynamic or creative UK judges will be in their interpreting of the Convention rights (Miles, 2001).

Notes

1 Contrast this with the State of Victoria's Infertility Treatment Act 1985. Under Section 8 (2) (3) (a) and (b) a person wishing to undergo treatment must either meet the requirement that ' … a doctor must be satisfied, on reasonable grounds, from an examination or from treatment he or she has carried out, that the woman is unlikely to become pregnant from an oocyte produced by her other than by a treatment procedure; or a doctor who has specialist qualifications in human genetics, must be satisfied, from an examination he or she has carried out, that if the woman became pregnant from an oocyte produced by her and sperm produced by her husband, a genetic abnormality or a disease might be transmitted to a person born as a result of a pregnancy'. (See also the constitutional challenge to this [McBain v State of Victoria, 2000].)

2 This limit has since been reduced to two (Human Fertilisation and Embryology Authority, 20001b).

References

All Party Lords and Commons Family and Child Protection Group (1983), *Evidence to the Warnock Committee*, House of Commons, London, 3 February.

Apoola, A., ten Hof, J. and Allan, P. (2001) 'Access to infertility investigations and treatment in couples infected with HIV: questionnaire survey', *British Medical Journal*, Vol. 323, p. 1285, 1 December.

Bennett, R. and Harris, J. (2001) 'The Welfare of the Child and Access to Infertility Treatment', *Journal of Fertility Counselling*, Vol. 8, No. 3, pp. 24-29.

BioNews (2000) 'Couple Demand Right to a Baby Girl', *BioNews*, No. 078 (2/10/00-8/10/00) www.progress.org.uk

BioNews (2001) 'Time to reconsider sex selection?' *BioNews*, No. 132 (5/11/01-11/11/01) www.progress.org.uk

Boseley, S., Ward, J. and Hartley Brewer, J. (2000) 'Human Rights Test for Fertility Rules', *The Guardian*, 14 February, p. 1.

Bottomley, V. (1990a) *Official Report*, House of Commons, 20 June, col. 1029.

Bottomley, V. (1990b) *Official Report*, House of Commons, 20 June, cols. 1029 and 1031.

Brazier M. (1999) 'Regulating the Reproduction Business?' *Medical Law Review*, No. 7, pp. 166-193.

Brown, C.J. (2000) 'Rationing Fertility Services in the NHS: The Patients' Viewpoint', *Human Fertility*, Vol. 3, pp. 268-270.

College of Health (1993) *Report of the National Survey of the Funding and Provision of Infertility Services*, College of Health, London.

Costa, G. (2001) 'Backdown on "Psychological Infertility"', *The Age*, 21 November, p. 3.

Council of Europe (1950) *Convention for the Protection of Human Rights and Fundamental Freedoms*, Council of Europe, Strasbourg.

Council of Europe (1989) *Human Artificial Procreation*, Council of Europe, Strasbourg.

Deech, R. (undated) *ART and the Family: Risk or Revival*, http://familycenter.byu.edu/marriageconf/media/deech.pdf

Department of Health and Social Security (1984) *Human Fertilisation and Embryology*, The Warnock Report, Cmnd 9414, HMSO, London.

Douglas, G. (1992) *Access to Assisted Reproduction: Legal and Other Criteria for Eligibility*, Report of a survey funded by the Nuffield Foundation, Cardiff Law School, Cardiff.

Douglas, G. (1993) 'Assisted Reproduction and the Welfare of the Child', in M. Freeman and B. Hepple (eds.), *Current Legal Problems*, Vol. 46, No. 2, Oxford, Oxford University Press, pp. 53-57.

Englert, Y. (1994) 'Artificial insemination of single women with donor semen. Artificial insemination with donor insemination: particular requests', *Human Reproduction*, Vol. 9, No. 11, pp. 1969-1977.

Human Fertilisation and Embryology Authority (2001a) *Code of Practice* (5th edn), Human Fertilisation and Embryology Authority, London.

Human Fertilisation and Embryology Authority (2001b) *HFEA reduces maximum number of embryos transferred in single IVF treatment from three to two* (press release), 8 August, Human Fertilisation and Embryology Authority, London.

J.R.M. v Netherlands (1993) Application No. 16944/90 European Commission of Human Rights, 8th February (reported in *Medical Law Review*, Artificial Insemination, Rights of Donor and Access to Treatment, Vol. 1, No. 3, pp. 404-7).

Kennelly, C. and Reisel, J. (1998) *Report of the Sixth National Survey of the Funding and Provision of Infertility Services*, College of Health, London.

Levitt M (2001) '"Not so Wrong that we are prepared to Threaten the Entire Service" – The Regulation of Reproductive Technologies in the UK', *Human Reproduction and Genetic Ethics*, Vol. 7, No. 2, pp. 45-51.

McBain v State of Victoria (2000) FCA 1009, 28 July.

Miles, G. (2001) 'Human Rights and the Regulation of Fertility Treatment', *Journal of Fertility Counselling*, Vol. 8, No. 2, pp. 21-28.

North West Lancashire Health Authority (undated) *Eligibility Criteria for Fertility Services*, North West Lancashire Health Authority, Wesham Nr Kirkham, Lancashire.

Peek, L. (2001) 'Children may get the right to trace sperm donor', *The Times*, 16 November, p. 13.

R v Human Fertilisation and Embryology Authority *ex parte* (1) Assisted Reproduction and Gynaecology Centre (2) H (2002).

R v Secretary of State for the Home Department *ex parte* Mellor (2001) EWCA Civ 472.

Rees v UK (1987) 9 EHRR 56.

Rogers, L. (2001) 'Woman to have Brother's IVF Child', *Sunday Times*, 26 August, pp. 1-2.

Saffron, L. (2001) 'Decision Making among Lesbians Wishing to get Pregnant', *Journal of Fertility Counselling*, Vol. 8, No. 2, pp. 31-36.

Soering v United Kingdom (1989) 11 EHRR 439 Stone, V. and Reisel, J. (1997) *Report of the Fifth National Survey of the Funding and Provision of Infertility Services*, College of Health, London.

Wiles, R. and Gordon, C. (1994) *Report of the Second National Survey of the Funding and Provision of Infertility Services*, College of Health, London.

Wiles, R. and Oddos, C. (1996) *Report of the Fourth National Survey of the Funding and Provision of Infertility Services*, College of Health, London.

Wiles, R. and Patel, H. (1995) *Report of the Third National Survey of the Funding and Provision of Infertility Services*, College of Health, London.

X, Y and Z v UK (1995) 75/1995/581/667.

PART III
ISSUES IN THE REGULATION OF ART

Chapter 14

Choosing A Child's Future? Reproductive Decision-Making And Preimplantation Genetic Diagnosis

Belinda Bennett

It is now more than 20 years since the world's first in-vitro fertilization (IVF) baby was born in England. Since that momentous occasion we have witnessed amazing advances in medical science. Medicine may not yet have a cure for infertility, but there is no doubt that for many infertile couples, the new reproductive technologies have been successful and have brought enormous joy into their lives.

Medical science has continued to develop, moving the horizons of the possible ever-further afield. In the past 25 years our map of reproductive medicine has been transformed. New features have appeared on that landscape, creating both areas of wonder and hazards to be wary of. If we walk across that landscape we see the older, and by now more familiar features such as cryopreservation of gametes and embryos, alongside more recent features such as widows who have used stored sperm to conceive their dead husband's child, and most recently, cloning and stem cell research. Around the edges of this new terrain and marking out 'no-go' zones within it, are the border patrol – the lawyers, ethicists and others who have sought to define the rules that apply in this new territory. This paper is about those border patrols. More specifically, this paper is about the borders that exist around preimplantation genetic diagnosis, and about where those borders might be placed.

Reproduction as contested territory

Reproduction is contested territory: it poses innumerable dilemmas and very few straightforward solutions, and it is something that affects everyone, whatever their ultimate reproductive choices, at a very personal level. To say that reproduction has become a contested site is not simply to acknowledge that the new reproductive technologies have posed an array of new ethical and legal dilemmas that defy easy solutions. It is also to acknowledge the social context of reproduction more broadly. It would be difficult to watch television and read newspapers without becoming aware of the ebbs and flows of public debate, law reform, cases, and advances in medical science and the intersections between them.

The moral significance of the embryo and control over its physical presence has taken on new dimensions in the past quarter of a century. On one level, the debate over the moral significance of the embryo is nothing new. It is after all, familiar territory from the debates over abortion in which the legal and ethical implications of terminating a pregnancy have been the ground on which countless battles have been played out in courtrooms and parliaments around the world (Petersen, 1993). Today we can hear echoes of those familiar debates in contemporary ones over reproductive research using human embryos, stem cell research and preimplantation genetic diagnosis.

What is clear is that the new reproductive technologies have challenged, and continue to challenge, our understandings of both the physical world and its cultural setting. Faced with the reality that medical science can achieve the previously unimaginable we remain unsure, as a community, about whether we really want to be able to achieve those things. The current debates over stem cell research that are taking place in Australia, Britain and the United States are an example of this. We may be on the threshold of some truly wondrous scientific advances, but serious concerns over the moral status of the human embryo and the implications of destroying embryos for research purposes remain for some sections of the community.

Even within feminist theory, there have been differing interpretations of the implications of the new reproductive technologies for women (Morgan, 1998). Some feminist scholars have expressed concern over the focus on high-tech treatments for infertility rather than its causes, and have argued that the new reproductive technologies reinforce social ideologies that women are unfulfilled unless they are mothers. Other feminist scholars however have argued that a decision to use assisted conception technologies must be seen as an expression of a woman's autonomous decision-making about her reproduction. Feminist scholars have acknowledged the need to avoid essentialist understandings of women's experiences by recognizing the importance of race, class, age, sexuality and infertility in women's lives (Spelman, 1988; Harris, 1990; Sandelowski, 1990; Lacey, 1998).

The law and ethics of preimplantation genetic diagnosis

In vitro fertilization techniques involve the fertilization of a human embryo outside the body of a woman. Once the embryo has reached a certain stage of early development it is then transferred back to the woman's body where it is hoped the embryo will implant and proceed to a pregnancy. For couples who are at risk of passing on an inherited condition if they conceive a child naturally, assisted conception technologies such as IVF provide one avenue for avoiding that outcome. A couple at risk of having a child with an inherited condition could use donated gametes or embryos so as to ensure that any child conceived is not the biological child of the affected parent. The other alternatives were for the couple not to have children, to have a child without genetic testing with the chance that the child could be born with a genetic disorder, or to use prenatal diagnosis during pregnancy to see if a fetus was affected by a genetic disorder and then to decide whether or not to terminate the pregnancy within the limits of the relevant abortion laws (HFEA/ACGT, 1999: para. 7).

Pre-implantation genetic diagnosis adds another option to the list of possibilities for parents faced with this situation by allowing embryos to be tested for the inheritable condition prior to being transferred to the woman's body (HFEA/ACGT, 1999: para. 2; Tasca and McClure, 1998: 7). This means that a couple at risk of passing on a inherited condition could both contribute gametes for the formation of the embryo (i.e. the embryos formed would be biologically related to both parents). The gametes could be fertilized using in-vitro fertilization techniques and then tested for the inheritable condition prior to being implanted. Only unaffected embryos would be transferred to the woman's body (HFEA/ACGT, 1999: para. 2; Tasca and McClure, 1998: 7). For sex-linked conditions, such as hemophilia, the sex of the embryo could be determined prior to transfer so that only those embryos that would develop into unaffected children would be transferred (HFEA/ACGT, 1999: para. 2; Tasca and McClure, 1998: 7).

Clearly the development of pre-implantation genetic diagnosis represents a significant advance in the treatment options that are available for couples who are at risk of passing on an inherited condition. Furthermore, with our increasing knowledge of genetics, it will be possible to screen embryos for known genetic mutations that could cause health problems later in life such as early-onset conditions (see for example, Verlinsky, Rechitsky, Verlinsky et al, 2002).

However, both sex selective technologies and the screening of embryos prior to implantation for genetic disorders are controversial and there is a need to determine the appropriate use of the technology. Public discussion on the appropriate use of the technology is an important aspect in the development of regulatory frameworks.[1] The reasons for the controversy are complex and involve overlapping concerns about the discarding of embryos as well as concerns over the potential eugenic implications of the technology and the possibility that the technology may be used for 'non-medical' reasons. Yet the importance of preimplantation genetic diagnosis for some infertile couples cannot be ignored: 'Some potential parents who carry a chromosomal rearrangement may never have achieved a viable pregnancy before requesting preimplantation genetic diagnosis if each previous conception resulted in a chromosomally unbalanced embryo which miscarried spontaneously' (Flinter, 2001: 1009).

In Australia, the regulation of preimplantation genetic diagnosis is governed by the laws and guidelines that regulate reproductive technology (Bennett, 2000). While Victoria, South Australia and Western Australia all have comprehensive legislation regulating the provision of assisted conception services,[2] preimplantation genetic diagnosis is only regulated in Victoria and Western Australia. In South Australia, the legislation does not address preimplantation genetic diagnosis but altering the genetic structure of an embryo is prohibited.[3] In other Australian states and territories, guidelines published by the National Health and Medical Research Council to provide ethical guidance on reproductive technology and the provision of assisted conception services (National Health and Medical Research Council, 1996) do not address the issue of preimplantation genetic diagnosis although research on embryos is regulated under the guidelines.

In Victoria, the *Infertility Treatment Act* 1995 prohibits the use of sex selection techniques with artificial insemination or in vitro fertilization unless 'it is necessary for the child to be of a particular sex so as to avoid the risk of transmission of a genetic abnormality or a disease to the child'.[4] The use of preimplantation genetic diagnosis in order to screen for embryos with genetic disorders or other abnormalities is not prohibited by the Victorian Act but altering the genetic constitution of a gamete or embryo is prohibited.[5]

In Western Australia, the *Human Reproductive Technology Act* 1991 prohibits causing or permitting 'a diagnostic procedure to be carried out upon or with an egg in the process of fertilization, or any embryo' unless the Code of Practice under the Act authorizes the procedure or the Western Australian Reproductive Technology Council has given its specific approval.[6] In addition, section 14(2) of the Act provides that the Human Reproductive Technology Council is not to give approval for research to be conducted or for a diagnostic procedure to be performed on or with an egg in the process of fertilization, or an embryo unless the Council is satisfied that 'the proposed research or procedure is intended to be therapeutic for that egg or embryo' and 'that existing scientific and medical knowledge indicates that no detrimental effect on the well-being of any egg in the process of fertilization or any embryo is likely thereby to occur'.

Disability and genetic testing

The development of preimplantation genetic diagnosis raises the ethical question of whether the technology is inherently eugenic. Similar issues arise in relation to the use of preimplantation genetic diagnosis for sex selection. Does the selective discarding of certain embryos represent a denigration of people with disabilities, or in the case of sex selection, of women? Is it wrong for prospective parents to choose to implant only those embryos that are not affected by particular genetic characteristics?

To a large degree the debates over preimplantation genetic diagnosis resonate with those about termination of pregnancy following prenatal diagnosis of fetal disability (Bennett, 2001; Jackson, 2000; Sheldon and Wilkinson, 2001). The legal context obviously differs – the difference between the regulation imposed by abortion laws and the regulation imposed by reproductive technology laws. For some people there would also be important moral differences between preimplantation genetic diagnosis, in which there is no pregnancy and no abortion of a developing fetus, and abortion which so clearly does involve terminating the life of fetus (Botkin, 1998: 20-21). However, since more embryos will be created than used, preimplantation genetic diagnosis may actually be seen by some as having greater ethical problems than abortion (Botkin, 1998: 21). What is clear is that the nature of reproductive autonomy and the significance of disability to reproductive decision-making are common territory for both preimplantation genetic diagnosis and prenatal diagnosis followed by abortion.

All prospective parents hope for a safe birth and a healthy child. It is understandable that prospective parents, when faced with a diagnosis of serious fetal disability following preimplantation genetic diagnosis, will be confronted with difficult choices. For couples who have chosen to use preimplantation genetic diagnosis to avoid having a child affected by a genetic condition, they are likely to already be aware of the condition and may already have an affected child.

Concern has been expressed over the potential harm that preimplantation genetic diagnosis and prenatal diagnosis may do to the rights and interests of individuals with disabilities (Asch, 1999). With genetic testing 'difference and abnormality are isolated by genetics in terms of "genetic potential", rather than in relation to the *ability of an individual to function in society*' (Taylor and Mykitiuk, 2001: 69). While the new genetics holds the potential for a new understanding of disability 'by highlighting the magnitude of human genetic variation and by demonstrating that everyone has genetic "abnormalities"' (Taylor and Mykitiuk, 2001: 69) this seems unlikely given the current focus of genetic research on identifying and treating genetic abnormalities (Taylor and Mykitiuk, 2001: 69). Furthermore, it has been argued that prenatal testing and abortion following a diagnosis of fetal disability is different from other forms of disability prevention, because it is not intended to 'prevent the disability or illness of a born or future human being but to prevent the birth of a human being who will have one of these undesired characteristics' (Asch, 1999: 1651):

> My moral opposition to prenatal testing and selective abortion flows from the conviction that life with disability is worthwhile and the belief that a just society must appreciate and nurture the lives of all people, whatever the endowments they receive in the natural lottery...people with disabilities do not merely take from others, they contribute as well - to families, to friends, to the economy. They contribute neither in spite of nor because of their disabilities, but because along with their disabilities come other characteristics of personality, talent and humanity that render people with disabilities full members of the human and moral community (Asch, 1999: 1652).

Asch argues that we need to move beyond our current limited views of disability and towards a society that is inclusive of disability.

Autonomy and the ethics of choice

A right to autonomous reproductive decision-making has been a key concern for women's group for decades. Major social and legal battles have been fought on the ground of abortion. In the context of abortion much of the debate has been about the rights of women to have access to safe and legal abortions, at least up to a certain stage of pregnancy. Restrictions on access to abortion, the content of abortion laws, and the difficulties women in rural and remote communities have accessing abortion services are all of prime importance in shaping the context of reproductive rights.

In contemporary Western societies such as Australia, the United States and the United Kingdom, respect for the autonomy of patients is regarded as being of prime importance in bioethical theory and has been recognized by the law. It is course, true to say that there are tensions between autonomy and beneficence that can rise to the surface in some health care contexts. However, on many occasions it is the patient's autonomy that trumps other concerns. In a variety of different contexts the law has acknowledged the importance of autonomy in health care and has upheld it (Jackson, 2000: 472-478).

It is important to see autonomy contextually for it is only when autonomy is analyzed in these terms that one can see that the decisions that women are often confronted with in a reproductive setting are made with reference to others. In recent years, there have been attempts to reconfigure the concept of autonomy and to move it beyond its traditional individualistic confines (Nedelsky, 1989; Stychin, 1998; Bennett, 1999). Jennifer Nedelsky for example has argued that autonomy must be seen in relational terms. She says 'Autonomy is a capacity that exists only in the context of social relations that support it and only in conjunction with the internal sense of being autonomous' (Nedelsky, 1989: 25)

A woman faced with a diagnosis that the fetus she is carrying has a serious abnormality that will result in serious disability may consider the implications of continuing the pregnancy and raising the child for other, existing children, and the implications for the family unit as a whole. Women are also likely to be the primary caregivers for children with disabilities caused by genetic conditions (Mahowald, 1996: 268-270). A decision about whether to proceed with a particular pregnancy is often a decision made with reference to other relationships:

> Because it is the fetus' relational qualities that women value – its role as the woman's developing child-to-be – women consider the rightness of their abortion decisions in terms of the decision's impact on themselves and the fetus as a unit. Just as pregnancy interconnects their present, so will birth intertwine them for the rest of their lives. When women consider the futures of their fetuses, their understanding of the dependency of babies and children ensures that they do not conceive of this future abstractly, but as being intertwined with their own (Cannold, 1998: 75).

Such decisions are profoundly relational (Gilligan, 1982; Cannold, 1998). 'Women's decisions about abortion are the same sorts of decisions they make about mothering, only with different outcomes' (Cannold, 1998: xvi). Given this relational dimension to reproductive decision-making it seems likely that decisions about preimplantation genetic diagnosis are also made in a relational context, with women making their decision with reference to the likely impact of their decision on other key individuals (including their future child) in their lives.

Individual decisions about reproduction are influenced by our social context, by our prevailing views about normalcy, and by the limits of our medical knowledge. It is important that we respect women's rights to utilize preimplantation genetic diagnosis or prenatal diagnosis to make reproductive decisions on the basis of the information received. To deny such rights would run counter to the arguments in favor of supporting a woman's right to choose in relation to abortion. In addition, 'forcing a woman to continue an unwanted pregnancy to term in order to achieve a shift in social attitudes or in scientific priorities would represent a marked exception to the current trend in medical law of treating patient autonomy as the overriding consideration' (Jackson, 2000: 481). However the disability critique of genetic testing reminds us of both the limitations of the new genetics, and our need to work towards a society that is more inclusive of disability, whether or not it is genetically caused.

Conclusion

A decision about a pregnancy is not an abstract decision but rather one about whether a particular pregnancy will be right for the woman concerned. Respect for autonomous decision-making in the context of preimplantation genetic diagnosis does not mean simply handing the issue over to individual rights which are conceptualized in an individualized and acontextual manner. Context and relationships *are* significant here.

The meaning of disability is complex, and becoming more so with our increasing knowledge of genetics. Preimplantation genetic diagnosis is likely to be utilized by relatively few couples because it is limited to those who are using IVF procedures. However preimplantation genetic diagnosis does provide important opportunities for debate and dialogue on the meanings of disability and reproductive choice in the context of the new genetics. Given how new these technologies are, it is vital for that dialogue to continue.

Notes

1 In the UK the Human Fertilisation and Embryology Authority and the Human Genetics Commission have recently reported on the results of a public consultation on preimplantation genetic diagnosis. See Human Fertilisation and Embryology Authority and Human Genetics Commission, 2001.

2 *Infertility Treatment Act* 1995 (Vic); *Reproductive Technology Act* 1988 (SA); *Human Reproductive Technology Act* 1991 (WA).

3 Reproductive Technology Code of Ethical Research Practice, s. 8 in *Reproductive Technology (Code of Ethical Research Practice) Regulations* 1995 (SA).
4 *Infertility Treatment Act* 1995 (Vic) s. 50.
5 *Infertility Treatment Act* 1995 (Vic) s. 39.
6 *Human Reproductive Technology Act* 1991 (WA) s. 7(1)(b).

References

Asch, A. (1999), 'Prenatal Diagnosis and Selective Abortion: A Challenge to Practice and Policy', *American Journal of Public Health*, 89, pp. 1649-1657.

Bennett, B. (1999), 'Posthumous Reproduction and the Meanings of Autonomy', *Melbourne University Law Review*, Vol. 23, pp. 286-307.

Bennett, B. (2000), 'Legal Regulation of Reproductive Technology in Australia', *Journal of Women's Health & Law*, Vol. 1(2), pp. 147-165.

Bennett, B. (2001), 'Prenatal Diagnosis, Genetics and Reproductive Decision-making', *Journal of Law and Medicine*, Vol. 9, pp. 28-40.

Botkin, J.R. (1998), 'Ethical Issues and Practical Problems in Preimplantation Genetic Diagnosis', *Journal of Law, Medicine & Ethics*, Vol. 26, pp. 17-28.

Cannold, L. (1998), *The Abortion Myth: Feminism, Morality and the Hard Choices Women Make*, Allen & Unwin, Sydney.

Flinter, F.A. (2001), 'Preimplantation Genetic Diagnosis: Needs to be Tightly Regulated', *British Medical Journal*, Vol. 322, pp. 1008-1009.

Gilligan, C. (1982), *In a Different Voice: Psychological Theory and Women's Development*, Harvard University Press, Cambridge, Massachusetts and London.

Harris, A. (1990), 'Race and Essentialism in Feminist Legal Theory', *Stanford Law Review*, Vol. 42, pp. 581-616.

HFEA/ACGT – see Human Fertilisation and Embryology Authority and Advisory Committee on Genetic Testing.

Human Fertilisation and Embryology Authority and Advisory Committee on Genetic Testing (1999), *Consultation Document on Preimplantation Genetic Diagnosis*, available at www.hfea.gov.uk.

Human Fertilisation and Embryology Authority and Human Genetics Commission (2001), *Outcome of the Public Consultation on Preimplantation Genetic Diagnosis*, available at www.hfea.gov.uk.

Jackson, E. (2000), 'Abortion, Autonomy and Prenatal Diagnosis', *Social & Legal Studies*, Vol. 9, pp. 467-494.

Lacey, L.J. (1998), '"O Wind, Remind Him That I Have No Child": Infertility and Feminist Jurisprudence', *Michigan Journal of Gender and Law*, Vol. 5, pp. 163-203.

Mahowald, M.B., Levinson, D. Cassel, C. et al (1996), 'The New Genetics and Women', *Milbank Quarterly*, Vol. 74, pp. 239-283.

Morgan, D. (1998), 'Frameworks of Analysis for Feminism's Accounts of Reproductive Technology' in Sheldon, S. and Thomson M. (eds), *Feminist Perspectives on Health Care Law*, Cavendish Publishing Ltd., London and Sydney, pp. 189-209.

National Health and Medical Research Council (1996), *Ethical Guidelines on Assisted Reproductive Technology*, Canberra.

Nedelsky, J. (1989), 'Reconceiving Autonomy: Sources, Thoughts and Possibilities', *Yale Journal of Law & Feminism*, Vol. 1, pp. 7-36.

Petersen, K.A. (1993), *Abortion Regimes*, Dartmouth Publishing Company Ltd., Aldershot.

Sandelowski, M. (1990), 'Fault Lines: Infertility and Imperiled Sisterhood', *Feminist Studies*, Vol. 16, pp. 33-51.

Sheldon, S. and Wilkinson, S. (2001), 'Termination of Pregnancy for Reason of Foetal Disability: Are There Grounds for a Special Exception in Law?', *Medical Law Review*, Vol. 9, pp. 85-109.

Spelman, E.V. (1988), *Inessential Woman: Problems of Exclusion in Feminist Thought*, Beacon Press, Boston.

Stychin, C.F. (1998), 'Body Talk: Rethinking Autonomy, Commodification and the Embodied Legal Self' in Sheldon, S. and Thomson, M. (eds) *Feminist Perspectives on Health Care Law*, Cavendish Publishing Ltd., London and Sydney, pp. 211-236.

Tasca, R.J. and McClure, M.E. (1998), 'The Emerging Technology and Application of Preimplantation Genetic Diagnosis', *Journal of Law, Medicine & Ethics*, Vol. 26, pp. 7-16.

Taylor, K. and Mykitiuk, R. (2001), 'Genetics, Normalcy and Disability', *ISUMA*, Vol. 2(3), pp. 65-71 available at: http://isuma.net

Verlinsky, Y., Rechitsky, S. and Verlinsky, O. et al, (2002), 'Preimplantation Diagnosis for Early-Onset Alzheimer Disease Caused by V717L Mutation', *JAMA*, Vol. 287, pp. 1018-1021.

Chapter 15

Stored Embryos And The Value Of Genetic Ties

Giuliana Fuscaldo

Introduction

How important is biological relatedness? Are genetic ties morally relevant in determining who is a mother or father, who incurs obligations towards children, who has claims over embryos and what children's entitlements are?

For many people these questions are easily answered intuitively. The desire for genetic offspring is often claimed to be a fundamental human need and a crucial self-defining experience. For some this is easily explained as the need to fulfill a strong biological urge. For others, genetic ties are important because they create shared family histories, a sense of continuity or a type of immortality. Many people find pleasure in the experience of shared family resemblance or traits and mannerisms. It is often claimed that genetically related parents and children form special attachments or that genetic ties give rise to special, immutable obligations and duties as exemplified by the old adage that *blood is thicker than water*.

However, it is also often claimed that in modern individualistic societies we have over-emphasized the importance of genetic relatedness. Anthropologists remind us that there exist cultural groups where genetic ties are not the sole or even the most important determinant of who has responsibility for children - where fosterage and surrogacy are not uncommon or where children are seen as communal responsibility.[1,2] Even in our own culture, it is argued, the many successful experience of couples who adopt or form blended families following divorce show that genetic ties are not the only determinant of parenthood.[3]

The development of IVF and the enormous growth in reproductive technologies largely reflect the importance in our society of biological parenthood and genetic kinship. However, the same technology has confused our understanding of what makes a parent and what significance we should attribute to passing on our genes. The advent of gamete donation and gestational surrogacy and the future prospects of collecting gametes from cadavers, parthenogenetically activating eggs to produce embryos, ectogenesis and reproductive cloning, necessitate a review of the significance of genetic ties.

This chapter investigates the value of genetic ties in relation to the decisions that people make about surplus frozen human embryos. I argue that our current attitudes and policies with regard to spare embryos are based on the unsupported assumption that genetic ties give rise to special moral obligations.[4]

Genes define parents

How important is it that the children we raise our genetically 'ours' and does passing on our genes make us parents?

As recently illustrated in a Melbourne press story covering paternity disputes,[5] for many people genetic ties are *very* important and do in fact define parenthood. The article, titled '*So you think you're the father?*' recounts the experiences of a growing number of individuals turning to DNA paternity testing, information that is then used in legal disputes over child custody and child support. The implications from this story are clear – that genes define parenthood and that in the absence of a genetic tie a 'putative father' is relieved of any duties and loses any parental claims, irrespective of the relationship that he and the child may have had. In other words, *my genes = my child and my responsibility.*

This story also illustrates that there is a sense in which we regard passing on our genes as 'having a child of our own'. Indeed it is the desire for just such a child that has led to the advent of IVF and the enormous growth in reproductive technologies. It is clear from the demand for these technologies that many individuals will go to great lengths to achieve biological and genetic parenthood.

There is also agreement that genetic ties are so strong that it is important to respect not just the desire to create such ties, but also the desire to avoid them. In recent legal battles over frozen embryos, great weight is accorded to the '*right not to be a parent*'[6] and typically it is this desire to avoid genetic parenthood that explains why hundreds of surplus human embryos are disposed of every year.

Surplus frozen embryos

A routine consequence of many in-vitro fertilization (IVF) procedures is that more embryos than patients can use for their own treatment are produced, these are cryopreserved for future use. However, when couples with frozen embryos complete or discontinue IVF treatment, they face the decision of what do with their surplus embryos. According to Victorian IVF laws, which came into effect on 1st January 1998,[7] human embryos can remain in frozen storage for a maximum of five years.

An exact picture of what happens to surplus embryos in Australia is hard to establish because this data is not collected by any central body, and because of the variation in or absence of legislation with regard to storage time limits for embryos. To date I am aware of only four Australian publications which give data on the fate of stored frozen embryos.[8] What each of these publications reveal is that very few people donate their unwanted embryos to others, preferring disposal or continued storage where possible. In Victoria, patients with surplus embryos reaching the five year storage limit face only two options:[9] to have the embryos disposed of or to donate them to another infertile couple. Two Victorian studies reveal that more than 80 per cent of patients in this position choose to dispose of their embryos while less than 10 percent donate their embryos to others.[10] It is interesting to note that the waiting list in Melbourne for donor embryos is from 18 months to 3 years with approximately 80 couples or women on the waiting list.[11] What should also be noted is that on average each couple has 3.7 embryos frozen (with a range of 1-14) so literally hundreds of embryos are disposed of every year from Melbourne clinics alone.[12]

Decisions about surplus embryos

Why do so few people donate their embryos?

In a 1998 study Kay Oke and colleagues, from Melbourne IVF sent questionnaires to all couples for whom the legal storage period was about to expire. Their study, presented at the Fertility Society of Australia in 1998, examined how patients reached the decision about whether to donate or dispose of embryos.[13] Their aims were to ascertain what couples wanted done with their embryos and to explore how they reached this decision. Among the questions they asked were '*how easy or difficult was this decisions*' and '*what factors influenced your decision*'. The following three figures give a sample of the responses collected by Oke; I have grouped these responses according to common themes.[14] The comments in Figure 15.1 reflect the notion that for some people, donating an embryo would be like giving away their child. Figure 15.2 reveals that some people choose not to donate their embryos because of concerns about the welfare of children raised in the absence of their genetic kin. While in Figure 15.3 the concerns expressed relate to legal aspects of donating embryos, such as problems associated with donor anonymity.[15]

> **If not donating, what factors influenced your decision?**
>
> "*It would be like giving a child away*"
>
> "*Couldn't cope with someone else raising our child*"
>
> "*My child not being in our family*"
>
> "*Our own flesh and blood being born to anyone other than ourselves, it would be too distressing*"
>
> (FSA 1998. K. Oke, K. Hammerberg and J. Blood)

Figure 15.1 Decisions about surplus embryos: own child

> **If not donating, what factors influenced your decision?**
>
> "*I would feel terrible for not giving the child its biological parents*"
>
> "*We could not bear the thought of our child being mistreated*"
>
> "*Unfair to have siblings in other families*"
>
> (FSA, 1998. K. Oke, K. Hammerberg and J. Blood)

Figure 15.2 Decisions about surplus embryos: child centered

If not donating, what factors influenced your decision?

"... *could trace us in future* ..."

"*may be able to access our details to seek us for any sort of emotional or financial support*"

"*rights of child to seek out biological parents* ..."

"... *the possibility of someone in twenty years time knocking on the door saying 'I'm your son/daughter!'*"

(FSA, 1998. K. Oke, K. Hammerberg and J. Blood)

Figure 15.3 Decisions about surplus embryos: legal uncertainty/anonymity

Assumptions

What the Oke study illustrates is that people feel very deeply about their embryos and equate donating embryos with becoming parents. The reasons given by couples deciding not to donate reveal a number of underlying assumptions about the significance of genetic parenthood:

1. That embryos belong to people who contribute their genes, in other words that my *genes = my child.*
2. that people are responsible for meeting the needs of children who develop from their genes or, *my child = my responsibility.*
3. that children should be raised by their genetic parents, in other words, that *genetic parents are best.*

Aims

My aim in this discussion is firstly to examine these assumptions. Is it reasonable to place so much value on genetic ties in the context of the fate of surplus frozen embryos? Secondly I will briefly discuss the policy implications of putting less value on genetic ties in the context of embryo donation.

But perhaps I need to begin by defending this question. It could be said that how people feel about their embryos is a private matter and not a matter for evaluation. Or that, people cannot help how they feel about their embryos because the need to pass on our genes and to nurture our genetic children is a strong biological urge or instinct. In response to the first objection, in this discussion I wish to raise the possibility of reflection on what has until recently been taken for granted. It may be that the way we view genetic parenthood now has arisen from historical, cultural or biological norms. In other words our ideas about what makes a parent emerged when reproduction was done the old-fashioned way. But things are not so simple now and reproductive technologies have opened up new possibilities for forming families and relationships. As evidence in the recent baby Jaycee case,[16] a child can have many different individuals in the role of genetic, gestational, rearing and legal parent. While we might agree that decisions about embryos are personal, we would also agree that these decisions should be guided by reason and should incorporate new insights or reflections.

With regard to the claim that genetic parenthood is an urge or instinct, it should be remembered that even if we are genetically predisposed to desire and feel responsible for our own genetic children (a highly contentious claim and one probably outside the realms of empirical investigation), this 'instinct' appears to be resistible or a feeling which we can *choose* to suppress or act on.

Claim 1: My genes = my child

'My child', what does it mean?

As we saw from the Oke study for many people embryos derived from their gametes represent *their child* and the reasons for not donating can be summarized as the desire not to be a parent or not to have a child of theirs "*out there somewhere*". Thus genetic ties are taken to define parenthood despite the fact that if donated these embryos would be raised by another family, and would entail no financial or legal obligations (save making available identifying information). But in what sense does "having my genes" constitute that the child that develops from a donated embryo is my child?

In exploring the idea of 'one's own child' Kenneth Alpern poses the following hypothetical:

> Suppose that, walking down the street, you were to discover that a baby in a stroller happened by chance to have exactly the same genetic makeup as you. Would this child be yours or you its parent? Would the discovery of this identity of genes necessarily or naturally engender in you deep affection for the child and concern for its well being?[17]

Alpern suggests that merely happening to have the same genes is not enough to constitute having a child of one's own, just as identical twins are not each other's child. What is missing? It may be that the genetic makeup of the child in the pram is only accidentally identical to my genetic makeup, but her genes did not come from my body. Nor were the child's genes derived from me in the way that genetic children are derived for their genetic parents' DNA "blue print".

My child = my own flesh and blood

To take each of these possibilities in turn, according to the "*my child = my own flesh and blood*" argument, what is important is that the child genes are made from materials produced by my body. But, exactly how much of my flesh and blood is my genetic child? As molecular biologist Lee Silver explains

> In the fertilized egg that eventually develops into a child, a mother deposits just a single copy of DNA for each of the 23 human chromosomes. A second set of 23 DNA molecules is deposited in this same egg by the genetic father. The information present is each of these 46 DNA molecules is copied over time into 100 million, million (100,000,000,000,000) *new* sets of DNA molecules that are placed into each new cell formed during fetal and child development. Each of these new DNA molecules is built from raw materials that are recovered form the food that the mother, and then the child consumes...[18]

Most of the DNA molecules that actually come from the mother or the father, if not all, disappear long before the child is born. At most only a tiny fraction of original DNA molecules from the parents survive in a few scattered cells among the millions of cells present in the child's body.[19] It seems implausible to suggest that this tiny fraction of DNA makes a child a physical continuation of its genetic parents.

My child = derived from my blueprint

What about the "*my blue print*" argument*?* Perhaps what is special about "*my own child*" is that her genes are derived from mine, they are a copy from my genetic blueprint. But, the problem here is that identical twins have the same blue-print. Are the children of my identical twin also my children because they are derived from the same genes? To borrow another hypothetical from Alpern, if it were possible for geneticist to map my entire genome and artificially create an oocyte carrying 23 chromosomes based on my genetic map would an embryo that develops from this egg be "my child"?[20] This does not seem to capture the essence of what it means to have our own child.

The cause and intention argument

But perhaps what makes "my child" special is that the fact that I intended and caused a child to exist whose genes are derived from my own. We *could* make a case that having intended and instigated the procedures that led to the existence of this child I am now responsible for it. But, if it is intention that makes the child "mine", does it matter whether or not the child is genetically mine? In other words, would this be any different to defining an adopted child as "my own" because I intended and instigated the adoption?

The combination argument

Maybe it is the combination of all three, my flesh and blood, my blueprint and my intention that defines a real parent and "my own child"? Although intuitively appealing, a recent technological innovation makes this approach to defining parenthood problematic.

In May 2000 it was confirmed that a baby carrying DNA from two women had been born. More than thirty such children have been born as a result of an infertility treatment designed to boost the chances of older women undergoing IVF.[21] Mitochondria (the cell's power-packs) from a younger woman's eggs were introduced into the eggs of an older woman, which were then fertilized with the sperm of the intended father. Because mitochondria have their own DNA, the resulting baby, (referred to from here on as 'baby Mic'), was genetically a mixture of the DNA from two women and a man. A Melbourne newspaper reported this development with the headline "One child two mothers" and suggested that technically baby Mic has three parents.[22] Who are Mic's real parents? Baby Mic is the product of bodily materials from two women and a man, that is, it is the "*flesh and blood*" of all three. Baby Mic's genes are derived from the genetic blueprint of two women and one man, and the women and men involved caused and intended baby Mic to exist. So, if genetic relatedness defines parenthood, then all three are the real parents, a possibility I will return to. But, this is an idea I think that many people would feel uncomfortable with because it does not help to determine who are the babies real parents in terms of who is responsible for baby Mic.

What this analysis has shown is that even though we might try discover who is a parent by aligning our ideas about parenting and families with tangible biological facts, genetic kinship is not so tangible.

Claim 2: My genes = my responsibility

Even if there were a way to make sense of the idea that my genes = my child this notion does not explain the claim that, *I am always responsible for any child with my genes*. It seems implausible to suggest that the woman who "donated" mitochondrial DNA is in any way responsible for the care and upbringing of baby Mic and yet it is precisely the idea that *genes = responsibility* that many people argue is what makes separating genetic and social parenting wrong. According to contemporary American philosopher James Nelson, genetic ties are not elective - that is, we can not simply choose to acknowledge them or sever them at will, because kinship comes with moral obligations. Nelson argues that because children are born in a helpless and needy condition and because genetic parents have caused such children to exist, they have an obligation to care for them.[23] This is the same reasoning that is used to argue that men should take responsibility for accidental pregnancies which they cause and to support paternity testing for assigning child welfare payments. However, there are some difficulties with the causation argument, how do we determine what or who is *the* cause of a child's existence? In the case of an IVF pregnancy is it the action of the genetic parents, the embryologist that inseminates the egg or the clinician that performs the embryo transfer? Perhaps the IVF baby would not have existed without the actions of its grandparents or great grandparents? Also it is not so simple explaining why causing something to happen entails that we are responsible for the consequences.[24]

In a modification of the causation argument law professor John Lawrence Hill reasons that it is the intended parents who are the first cause of a procreative relationship.

> …while all the players in the procreative arrangement are necessary in bringing a child into the world, *the child would not have been born but for the efforts of the intended parents*[25] [original emphasis].

Hill argues that what is essential in determining competing claims over children is not the biological tie between a parent or a child but the "preconception intention" to have a child. He maintains that although it might seem peculiar to determine parental status on the basis of mental states and not tangible biological fact, there is a legal precedent for the relevance of mental states in other areas of law including contract torts and criminal law.[26]

But if parental obligations are determined according to intent then why do we pursue and label recalcitrant men who never intended to be fathers who refuse to pay child support? What would be wrong with a woman who became pregnant accidentally, simply abandoning her children on the basis that she never intended to be a mother?

A less problematic and widely accepted view of causation and moral accountability is that *we are morally accountable for the intended and unintended reasonably foreseeable consequences of our actions.*[27,28] In other words if we cause a child to exist, it is foreseeable even if not intended, that this child will not survive without care. To the extent that donating an embryo causes a child to exist, it could be said that the genetic progenitors of a donated embryo have moral obligations towards the child that results from the donation because this result was reasonably foreseeable.

However, there are a number of different ways of meeting this obligation. The genetic parents of the embryo could look after the resultant child themselves, they could transfer or discharge their obligations to others, or they could share the responsibilities of parenting with other willing individuals. If embryo donors can reasonably foresee that a child that develops from their donation will have its needs met, then they have met their obligations.

In the case of gamete and embryo donors it seems reasonably foreseeable, given the long waiting lists, that such donations will cause a child to exist for whom there are many people willing and able to take on parental responsibilities.

What this analysis has shown is that even if Claim 2, *my genes = my responsibility*, were true, it does not follow that my genes = my non-transferable responsibility. Despite the qualms of IVF patients with surplus embryos, donating their embryos to others that intend to parent does not amount to failing in their moral duties.

Claim 3: Genetic parents are best

It could be that what is feared by people in a position to donate embryos, is that something will be lost in the transfer of responsibility, that children's needs might not be met, that they might suffer in some way or exist in a less than ideal state. In other words it may prove difficult to ensure that the consequences of our actions with regard to embryo donation are not such that they attach moral blame.

As revealed in the Oke study, some of the respondents believed it would be hard on or unfair to children to separate them from their genetic heritage or their genetic siblings. Others expressed concern that the home they would be raised in might be less than ideal. Which brings us to the claim that genetic parents are best because children raised by other than their biological parents are somehow harmed.

There is much current debate about the need for a child to know and be raised by his or her genetic parents.[29] Among the advantages of combining genetic and social parenting it is claimed that genetic parents form closer emotional attachment to children who carry their genes or that children, who grow up in traditional genetically related families achiever better outcomes. Evidence in support of this conclusion has been presented against a wide range of criteria including emotional and psychological well being, educational outcomes, involvement with drugs, work patterns, delinquent or criminal behavior, risk of child abuse, sense of identity, et cetera.[30] What can be deduced from the evidence is that there are a large number of variables which could affect children's welfare including parental socio-economic status, availability and extent of social support, genetic factors, and the emotional and psychological well being of the rearing parents.

It is outside the scope of this paper to give a detailed evaluation of all the data for and against the claims that children do better when raised by their genetic parents and, given the complex number and combinations of possible variables, it is doubtful that such claims can ever be satisfactorily tested empirically. Suffice it to say that there is room to doubt the claim that genetic parents best can meet children's needs. Nor in fact, is it necessary to prove this claim for the purpose of answering the question posed. The question is, are genetic ties so important to children's welfare that it is better to destroy embryos than to donate them to other infertile people?

In response to the claims that it is wrong to create children who will be cut off from their genetic heritage or suffer the same genealogical confusion experienced by some adoptees - if this is the concern of people who choose not to donate - then perhaps the problem with the way embryo donation is practiced in this country is that it precludes the possibility of acknowledging both the genetic and social parents. In other words, even if it is true that *my genes = my child* it does not follow that my genes = no one else's child or that children can have only two parents. Perhaps donor anonymity with also cease to be any issue if we acknowledge that genetic parents are just one type of parent and genetic heritage only one way of understanding our identity and history.

Policy implications

In this discussion I explored the importance of genetic ties by examining the reasons that people give for preferring their surplus embryos to be discarded rather than donated to other couples. I have argued against the assumptions underlying these reasons that parenthood should be genetically defined and carries with it special duties and that children are harmed if raised in the absence of their genetic kin. Further, I suggest that parental duties can be transferred or shared and that even if we accept the claim that duties stem from genetic relatedness, it does not follow that embryo donors fail in their parental duties. In terms of our attitudes and policies regarding surplus human embryos I suggest that open or directed embryo donation the way it is practiced in some states of America[31] might ease some of the anxiety that couples feel about 'their' child's future. Such programs permit negotiated levels of interaction between donors and recipients and acknowledge both the contributions of social and genetic parents. There is no requirement that a child have two and only two parents.

If genetic ties are not so important, then, given the long waiting lists, the pain of infertility and the reduced costs associated with utilizing surplus frozen embryos, and given that we can ameliorate the anxiety felt by those faced with the prospect of donating, then we should encourage embryo donation in the same way that we currently promote organ donation.

Notes

1 Sault, N.(2000) Many mothers, many fathers: the meaning of parenting around the world. Santa Clara University Law Review. 36:2. [Online accessed 23/07/00] URL: http://www.scu.edu/SCU/Centers/Ethics/publications/other/lawrview/manymothers.html
2 Stone, L. (1997), *Kinship and Gender*. Westview Press, Harper-Collins, Boulder.
3 Bartholet, E. (1993), *Family Bonds Adoption and the Politics of Parenting*. Houghton Mifflin, Boston.
4 I am indebted to Dr Lynn Gillam and Professor Julian Savalescu for helpful discussion in the preparation of this paper and to Kay Oke for access to raw data.
5 Bearup, G. "So you think you're the father…". *The Age Magazine, Good Weekend*, 3 November, pp. 16-21.
6 For example Davis v Davis (1990) and Kass v Kass (1998) discussed in Robertson, J.A. (1994) *Children of Choice*. Princeton University Press, Princeton.
7 *Infertility Treatment Act* (1995) Victoria, Act No. 63/1995.
8 Darlington, N. and Matson, P. (1999) The fate of cryopreserved human embryos approaching their legal limit of storage within a West Australian in-vitro fertilization clinic. *Human Reproduction*. Vol. 14, No. 9, pp.2343-2344.
9 Although in some circumstances the Infertility Treatment Authority (the statutory body which administers Victoria's IVF legislation) will grant an extension to couples applying to store their embryos for longer than five years.
10 Breheny, op cit., Oke, op cit.
11 Oke, K. (2001). Personal communication.
12 Breheny, S. (2001). Personal communication.
13 Oke, op cit.
14 I am indebted to Kay Oke for access to raw data and questionnaire responses.
15 Victorian IVF laws require that gamete and embryo donors provide identifying information which can be made available to children born through such procedures on turning eighteen years of age.
16 Vorzimer, A.W., O'Hara, M.D. and Meyers, L.S. (1998) 'Buzzanca v Buzzanca: The Ruling and Ramifications'. *Journal of Assisted Reproductive Law*. Vol. 1, No.1. [Online accessed 18/12/00] URL: http://www.surrogatelaw.org/jaycee2.htm. This case involved the custody of Jaycee, a child conceived following anonymous embryo donation and gestational surrogacy. Technically Jaycee has five parents, a genetic, gestational and intended mother and a genetic and legal father.
17 Alpern, K.D. (1992) 'Genetic puzzles and stork stories: on the meaning and significance of having children', in K.D. Alpern (ed.), *The Ethics of Reproductive Technology*. Oxford University Press, Oxford, p. 160.
18 Silver, L.M. and Silver, S.R. (1998) 'Confused heritage and the absurdity of genetic ownership'. *Harvard Journal of Law and Technology*, Vol. 11, No. 3, p. 600.
19 Silver, op cit.
20 Alpern, op cit.
21 Barritt, J.A., Brenner, C.A., Malter, H.E. and Cohen, J. (2001) 'Mitochondria in human offspring derived from ooplasmic transplantation'. *Human Reproduction.* Vol. 16, No. 3, pp. 515-516.
22 Heighfield, R. and Heinrichs, P. 'World's first genetically altered babies'. Sunday Age, 6 May, 2001, p. 1.
23 Nelson, J.L. (2000) Reproductive ethics and the family. New Zealand Journal of Bioethics, June, pp. 4-10; Nelson, J.L. (1991) 'Parental Obligations and the ethics of surrogacy: a causal perspective'. *Public Affairs Quarterly*, Vol. 5, No.1, pp. 49-61.

24 See Kagan, S. (1988) 'Causation and responsibility'. *American Philosophical Quarterly*, Vol. 25, No. 4, pp. 293-302.
25 Hill, J.L. (1991) 'What does it mean to be a "parent"? The claims of biology as the basis for parental rights'. *New York University Law Review*, Vol. 66, May, p. 414.
26 Hill, p. 415. op cit.
27 Coghlan, M. (1990) 'Essential aims and unavoidable responsibilities'. *Bioethics*, Vol. 4, No. 1.
28 Cagan, S. (1989) *The Limits of Morality*. Oxford University Press, New York.
29 Turner, A.J, and Coyle, A. (2000) 'What does it mean to be a donor offspring? The identity of adults conceived by donor insemination and the implications for counselling and therapy'. *Human Reproduction*. Vol. 14, No. 9, pp. 2041-2051; Frith. L (2001) 'Gamete donation and anonymity'. *Human Reproduction*. Vol. 16, No. 5, pp. 818-824.
30 For a summary of research examining how children are affected by not living with two biological parents see Cherlin, A.J. 'Going to extremes: Family structure, children's well-being, and social science'. *Demography*. Vol. 36, No. 4, pp. 421-428.
31 For example of open embryo adoption programs see Creating Families at http://www.eggdonorfertilitybank.com and Snowflakes Embryo Adoption Program at http://www.snowflakes.org/FAQs.htm.

Chapter 16

The Public Interest In Embryos

Julian Savulescu

Role play

"How do people have children in your country?", the Stranger asked.

"Why, a man meets a woman, they fall in love, marry, have sexual intercourse, which usually results in a child. Most people have 2 or 3 children", I said

"And what happens then?"

"What do you mean?"

"What happens to the children after they are born?", the Stranger asked.

"Well, they are fed, loved, and protected, until they are of an age that they can go to school, and then to work", I replied.

"You mean these children are brought up by their genetic parents and live their childhood with them?", the Stranger asked.

"Yes, of course."

"How strange things are in your land", said the Stranger.

"Why, what do you mean?", I asked.

"Things are very different where I live. There are histories which tell of a mode of living like yours, but this was rejected in our country many centuries ago. In my land, when a couple wants to have a baby, each partner donates gametes. These are used to produce a number of embryos. These embryos are tested to ensure that there are no genetic diseases or predispositions to disease. Once it is ascertained that the embryos are healthy, they go to a distribution area, called the Fertility Center. From there, embryos are randomly distributed to couples wanting a baby. Records are kept about which embryo is from which couple only until the distribution, to ensure that no couple receives their own genetic offspring by chance. Everyone receives the genetic offspring of someone else. All records are then destroyed. No one ever knows who was their genetic progenitor."

"What a bizarre system", I said.

"On the contrary, we have found it to be superior to that which nature has shouldered you with. Since this system applies to everyone, everyone accepts it. We recognized early that genetic privilege tends to translate into social privilege, because after all, those genetic traits were selected by millions of years of evolution because they tended to confer social advantages which in turn conferred reproductive or survival advantages. So our system ensures that those who are genetically disadvantaged will have a higher chance of being reared in socially advantageous circumstances. Those who happen to be placed in disadvantaged social circumstances tend to have more genetic advantages, and tend to rise out of their social circumstances. It is rational and egalitarian. The social lottery corrects for the effects of the genetic lottery."

"It is much easier to deal with infertility in our society. People do not struggle to have their own genetically-related child because no one does. Spare embryos are given to infertile couples. Since all couples are used to giving up their gametes, they find this arrangement very satisfactory. Moreover, every couple enjoys the bonding which occurs during pregnancy."

"But how can such a system work? Don't people reject the child they are given and seek out their own? Don't the women who must give up their child suffer great sadness when it is taken from them. Aren't children neglected?", I asked.

"On the contrary, the system works very well. Children do better. Parents love them as individuals themselves, and not as some sort of replica of themselves. Parental narcissism has disappeared from our culture. The demands of genetic expectation do not exist. Our people love their children, and they no longer care about blood ties or genetic relations. These, they see, are the outdated currency of brute biological evolution. We are free and make rational decisions about what is best for us. Last year, our people were given the choice to return to the old system. No one voted for change."

This example raises questions about how important genetic relatedness is and how bad it is to have one's own genetically related child reared by someone else. These questions are at the heart of a number of important issues in assisted reproduction. One is the fate of frozen embryos. It is common place in assisted reproduction for couples to produce more embryos than they need to complete their family. These are called "excess" or "spare" embryos. These frozen embryos can remain in storage, be destroyed, be donated for research (in some places) or be donated to other infertile couples who cannot produce embryos. Infertile couples entering IVF are reported to show great solidarity – claiming they will donate spare embryos to other infertile couples. However, this only rarely happens. Once their family is completed, their altruism and solidarity apparently evaporates. Couples frequently choose to have their embryos either remain frozen in storage or be destroyed, rather than donate them to other infertile couples.

Is this wrong? Is it wrong to destroy one's embryo rather than donate it to an infertile couple?

Many people believe that the fate of one's embryos must remain a personal decision and is one of the most private decisions a person can make. These people believe that it is purely a matter of personal choice whether an embryo is destroyed or donated to another couple. The very idea of "donation" captures the sense that it is supererogatory, like choosing to donate one's kidney to a stranger.

I will argue that not only is it selfish and undesirable, it is wrong not to donate one's spare embryos to an infertile couple. But "wrong" can have several implications. It can imply that couples should be advised to donate their embryos. Or it might imply their embryos should be forcibly removed and given to infertile couples. I will argue for an intermediate position between these weak and strong senses of "wrong". I will argue that couples should be informed that their spare embryos will be given infertile couples unless they register a strong objection.

This opt out system is similar to that which has been proposed for organs and tissues in general. Embryos spare from IVF treatment, like organs and tissues after a person's death, should be made available for the benefit of others, unless there is a strong objection.

What views have been expressed on this kind of issue? The law has considered a case which yields some light on the ethical issues which are involved.

The case of Davis v. Davis and the fate of unwanted embryos

Davis v. Davis concerned the disposition of seven frozen embryos, produced by a couple who subsequently went through divorce proceedings. Initially the wife, Mary Sue, sought control of the embryos because she wished to attempt a pregnancy. Both remarried, and Mary Sue no longer wanted to use the embryos herself. She did, however, wish to donate the embryos to a childless couple, while Junior Davis staunchly opposed this move, preferring to destroy the embryos.

The Supreme Court identified the essential dispute as conflict between two rights: the right to procreate and the right not to procreate.[1]

While mindful of Mary Sue's greater labor in producing eggs for the IVF process, the Supreme Court found that she and Junior Davis should be considered 'entirely equivalent gamete-providers'. This was decided after balancing the couple's respective experiences of gamete donation with the 'joys of parenthood or the relative anguish of a lifetime of unwanted parenthood'. The court recognized that any process resulting in parenthood was dramatic. The court concluded 'that an interest in avoiding genetic parenthood can be significant enough to trigger the protection afforded to all other aspects of parenthood'. The court recognized further that in the event of strangers successfully gestating another's embryos, the gamete-providers become 'genetic' parents, bringing about a 'profound impact'.

The Supreme Court found in favor of Junior Davis, identifying the burdens of the imposition of unwanted parenthood to be overriding. The likely impact of unwanted parenthood was established by looking to Davis's particular circumstances, including his 'boyhood experiences', and his relationships with both of his parents.[2] His fear for any psychological consequences suffered by the future child was taken into account, as was his concern that any couple in receipt of the embryo might divorce, leaving the child in a single parent setting. These considerations were balanced against Mary Sue's desire to donate the embryos. The court was not willing to consider the interests of a recipient couple to whom Mary Sue wished to donate the embryos.

The court in this case claimed there was a profound impact in becoming a genetic parent. Genetic relatedness or blood ties are accorded great weight in society. But why should they be considered to be of such value? How specifically is it bad to have provided genetic material to a child whom one does not raise?

It is clear that having access to spare frozen embryos would be of great value to society and to the infertile couples who cannot produce their own embryos and, arguably, to the child who results. Since it is not routine practice to pass on these embryos to infertile couples, or even to recommend it, there must be something very bad about spare embryos going to infertile couples against the wishes of the couple who produced them.

What is bad about spare embryos going to other infertile couples?

1. Interference in autonomy

Carson Strong[3] claims that to interfere with the freedom not to have children is to treat persons as mere means rather than ends in themselves. He claims that depriving persons of this decision constitutes a failure of respect, and interferes with their self-determination and self-fulfillment. According to Strong's view, the fact that blood ties are valued by people provides a reason for action.

This view is mistaken. We can value what is not of value or valuable. The fact that people autonomously desire to control the fate of their genes does not by itself establish that they have a good reason to so act. History is replete with examples of people desiring to do what there is no good reason to do: e.g. imprisoning or beating up homosexuals, keeping slaves, and so on.

2. *Children have better lives if raised by a genetically related parent than if raised by a non-genetically related parent*

Some people may want to have a genetically related child or value genetic relatedness because they believe either that the resulting child will have similar characteristics to them, such as appearance, intelligence, personality or temperament. These people may believe that children who are not genetically related to their parents will be thoroughly different to their social parents, and this will have bad consequences for child rearing. Call this the value of phenotypic similarity thesis.

The starkest example is a black couple trying to raise a white child in a black neighborhood. It might be thought that having a number of children that look like each other and the parents provides a greater sense of group cohesion and a sense of belonging.

Even if there is value to resemblance and similarity in parents and their children, that is, phenotypic similarity, this view, when widely applied, is based on a crude genetic determinism. The degree to which a given trait is genetically determined varies. Consider physical appearance. One study shows that, despite widespread conviction to the contrary, children look no more like their parents than any other adults.[4]

Intelligence is estimated to be 50% heritable, but this is a measure of variance between populations. It is true that a genetically related child has a greater probability of resembling one or another parent in certain respects, at least for those traits under the control of a simple genetic mechanism. However, more complex traits like personality are polygenic, and therefore subject to 'regression to the mean'. This means that a child borne by exceptional parents is likely to be unexceptional due to the recombination of an enormous amount of genetic material. A child is not just a 'mix' of parents' genes, but a 'recombination' such that the child has new genes, differing from both parents, and generative of a novel phenotype.

To summarize, the current state of knowledge suggests a genetic contribution to many traits, but the magnitude of this is not clear. There is reason to question the extent that a child will be phenotypically similar to his or her parents. But let us assume that children are more phenotypically similar to their genetic parents than other people. Does this imply that the value of phenotypic similarity thesis is true?

No. It is simply not clear whether being phenotypically similar to one's parents is important to having a good childhood. It is important to remember that between 5-20% of children are not the children of their social father, and neither he nor they know that. Non-paternity is common and families appear to function in the face of non-paternity.

Moreover, there are valuable lessons to be learnt from the social practice of adoption. A large Swedish study[5] followed adopted children from an early age through to adulthood. Results of this study were positive and showed the children coped well. Another large American study[6] of 715 children adopted as infants confirmed this. This study showed that adoptees' major reason was interest in parent's physical appearance, while the fifth was curiosity as to why they were adopted. Significantly, the desire to form some sort of relationship with birth parents did not rate a mention, nor did the desire for further understanding of self-identity. Another study found that adopted children are likely to become settled in the community.[7]

It is simply not clear that phenotypic similarity to one's social parents is necessary for a child to have a happy and good childhood. It might be objected that, phenotypic similarities or dissimilarities aside, simply being genetically unrelated to one's parents may be psychologically disturbing, and make a person's life go badly.

Empirical study of children born through artificial insemination would answer this question. While there are vocal malcontents with the practice of AID, and some people who claim to have been damaged by being the child of AID, it also appears that sensitive disclosure early in childhood facilitates adjustment. It is unlikely that being genetically unrelated to one's parents is going to severely compromise a person's life, except in exceptional circumstances.

Most importantly, even if it were true that children do better if raised by their genetic parents, it must be remembered that the alternative for these embryos to being raised by genetically unrelated parents is destruction. No matter how bad the life of a white child in a black household, it is unlikely to be so bad as to justify the claim that it would have been better if that child had never existed. Put simply, donation must be, in all but the most exceptional circumstances, better for the children who come into existence than embryo destruction.

3. It is counter-evolutionary

Some authors have suggested that the desire to beget progeny genetically related to oneself is important because it is fundamental to human nature as a result of evolution. The study previously mentioned which examined physical resemblance also found that at one year of age, children do resemble their fathers.[8] The proposed evolutionary explanation suggested that such resemblance provided fathers with a way of confirming paternity, justifying their nurturance and care.[9] Obviously, women are assured of which progeny is theirs, and do not need the same marker of parenthood in order to secure parental investment. Physical resemblance at a young age has been important because it provided men with a reason to care for their children. Perhaps children who did not resemble their father were abandoned, or cared for unequally, selecting for those that did.

Even if valuing genetic relatedness were selected by evolution, it is unclear why we should continue to value it the absence of selective pressures. There are many cases in which what is instinctive or natural no longer serves practical purposes or moral progress. The ability to reflect upon and control our instinctual drives distinguishes humans from other animals. Just because it is natural does not imply it is good. For example, it has long been postulated that the male monopoly of violence can be explained by evolutionary pressure on males to be aggressive. Nonetheless, few would deny that this is a trait, however natural or valuable in evolutionary terms, that has ceased to be valuable in our current social context.

4. Value of shared ancestry/history

Many people desire genetically related children because of deep-seated feelings about shared ancestry and family history. When we think about the constitution of shared ancestral information, however, it becomes clear that it has little to do with genetic material. What is important about our sense of ancestry, and family connectedness between and within generations are family stories, knowledge about our relations and their history. These things do not cease to become important to, say, an adopted child of a family. "Grandma" is perhaps important, not merely because you have her nose, but because of what she taught your mother, how she loved your grandfather and how she is the source of your inspiration in certain subjects.[10]

That is, what sort of person she was. Perhaps, the more common adopted children become, or the more people reflect upon what is important in a family, the less value we will place on shared genetic material. Consider a parallel with friendship. Friends are not genetically related. Yet they can have as intimate and deep a relationship as genetically related relatives. They can have a shared history and understanding, and all those features we take to be important in family traditions.

5. Blood ties give rise to special obligations

We cannot care equally and deeply about everyone. Blood ties are one way of dividing whom we should care most about from whom we should care less about. From a social point of view, the value of the family unit as mode of efficient care and nurturing, and the usual prominent place genetic links play in families, give great value to genetic relatedness.

But families are changing. Many families now have members who are not genetically related. There is no good reason to give greater weight to genetic links rather than social links. The divider of genetic links is arbitrary or at least one given to us by nature. We are able to care deeply and effectively for our friends, despite their not being related genetically to us. It is up to us to decide how we carve up our pattern of concern.

In conclusion, genetic relatedness is accorded great value in our society and many important social institutions and laws are predicated on its value. However, on closer scrutiny, it is difficult to justify the value accorded to genetic relatedness. This is not to say it is not of value at all. It is only to say that it may be overvalued.

This is all well and good but surely you have missed the issue here, someone is liable to object – what we are talking about is a couple's embryo, their own flesh and blood metaphorically speaking. There is something special about that. And it is because of that specialness that couples retain the right to decide what happens to their embryos. Because embryos are special to people, they have a strong autonomy based right to determine what happens to them. Perhaps it could even be said that they own that embryo and so have the right to dispose of it as they wish.

There are two possible ontological views of the embryo. Firstly, some people hold that the embryo is a person. If the embryo is a person, then it is wrong to kill it. Parents do not own their own children and parents do not have the right to kill them, even if they do not want them. If the embryo is a person, there is a very strong reason to require donation to other infertile couples.

What if the embryo is not a person but merely a bunch of cells or a piece of tissue? Surely people have the right to determine what happens to their own tissue. Indeed, there is great enthusiasm for giving people greater ownership rights over tissue, and power to control its fate once it is outside their body, as the Alder Hey scandal most acutely illustrates.

There are two responses. Firstly, in law, people do not own their own tissue or body parts. "No property in the body." But this of course could change. There is increasing excitement over giving people the right to control what happens to bits of their body and their DNA. We might even create a property right over one's own tissues. This view provides the strongest support for allowing couples to destroy rather than donate their embryos.

I am inclined to think that granting property rights or rights of self-determination over bits of one's body will not resolve the problem. The reason is that even if we have property in tissue, there seem to be other overriding moral considerations If we have property in tissue, this implies that we should be allowed to destroy our organs and tissues after our death even when these would save people's life. But this is unethical. It is clearly unethical to prefer to have one's kidney burnt than used to save a life. We have a duty of easy rescue to donate our organs and tissues, whether or not we have a general property right over them. Property rights might entitle a person to sell their organs, but there is a public interest in organs and tissues that should be overriding if we have not made appropriate provision for their fate after our death.

In a similar way, I believe that couples producing an excess of embryos are under an obligation: the duty of easy rescue. According to this duty, there are two necessary conditions for a State intervention, I, in a person A's behavior or life to be justifiable for the benefit of some other person B:

> the harm to A of I is below some acceptable threshold, and
> the harm to B of not-I is great.[11]

In this case, the harm to fertile couples of having the embryos go to an infertile couple is minimal. The benefit to the infertile couple is great. At very least, we have good reason to move to an opt out system of spare embryo donation. Of course we should first try to persuade couples to donate their embryos. But if this is unsuccessful, we should move to an opt out system for embryo "donation", just as we should move to an opt out system of organ donation.

What are the implications of reassessing the value of genetic relatedness?

Lessening our attachment to genetic relatedness or blood ties would have a number of important implications:

Firstly, unwanted eggs/embryos would be more likely to be donated. This is of considerable importance when we consider the great happiness a child brings to a childless couple. If we include the value of the life produced, as I believe we should, there is strong reason to use spare embryos for other infertile couples who cannot produce embyros.

Secondly, sperm donation would increase. This would have similar valuable consequences.

Thirdly, adoption would become a more valued way of having children. If people could be encouraged to adopt, this might facilitate a more rational approach to overpopulation.

Fourthly, couples might be more prepared to adopt out their own children. This would be very useful in one context. Although there are a number of prenatal genetic tests (e.g. Down syndrome screening) designed to pick up disabilities before birth, no combination of tests can detect all abnormalities prior to birth. One strategy to avoid having a disabled child which is not routinely mentioned to couples should be to examine the newborn child and to adopt out a disabled child. The major barrier to this is a sense of commitment to one's genetic offspring.

Fifthly, reducing our commitment to genetic relatedness opens the way for widespread genetic manipulations which introduce genes not present in the parents. This may be important in the future as gene therapy becomes a reality

Lastly, devaluing genetic relatedness would weaken the argument that women have the right to destroy their babies rather than give them up for adoption or artificial gestation (or ectogenesis).

Notes

I wish to thank Jacinta Kerin for invaluable research assistance in developing this paper.

1 It has been argued in feminist ethical analysis that these rights should not be considered equal, since a negative right cannot dictate the lengths that should be taken in order to fulfil a positive right. Interestingly enough, the eventual decision conforms to this point.

2 The detail into which this court was prepared to delve in order to ascertain Junior Davis' specific concerns is quite outstanding, especially given the ignorance of the context in

which women's reproductive decisions are made characteristic of so many legal decisions.

3 Carson Strong, *Ethics in reproductive and perinatal medicine: A new framework*, London: Yale University Press, 1997, p. 39.

4 Nicholas J. S. Christenfeld, Emily A. Hill, 'Whose baby are you?' *Nature*, Vol. 378, No. 14, Dec. 1995, p. 669. This study was comprised of 24 families, half of whom provided pictures of male children, the other half of female children, all at three different sets of ages. 122 subjects compared each stimulus picture with three possible parents, one of whom was the true match.

5 Bohman and Sigvardsson, 'Outcome in Adoption: Lessons from Longitudinal Studies' in David M. Brodinsky and Marshall D. Schechter (eds) *The Psychology of Adoption*, New York: Oxford University Press, 1990.

6 Benson, Sharma and Rochlkepartain, *Growing Up Adopted: A Portrait of Adolescents and Their Families*, Minneapolis: Search Institute, 1994. Study of 715 confirms positive outlook for children adopted as infants before 15 months. Reasons given to find birth parents: 1) to see what they look like, 94%, 2) to tell them I'm happy, 80%, 3) to tell them I'm ok, 76%, 4) to tell them I'm glad to be alive, 73% and 5) to know why adopted, 72%. Note that interest to form a relationship with birth parents, or develop self-identity not in top five reasons.

7 Triseliotis and Hill, 'Contrasting Adoption, Foster Care, and Residential Rearing' in Brodinsky and Schechter, op. cit.

8 *Nature*, p. 699, op. cit.

9 Two physical bases for this resemblance were proposed. The first was that the variability of babies' faces is more similar to the variability of fathers' faces than of mothers' (i.e. features salient on men are more visible on one-year-olds). The second was that a baby's facial morphology might be more determined by the father than the mother due to possible genomic imprinting. Note that neither of these explanations need necessarily be linked with the evolutionary hypothesis mentioned in the text.

10 An adopted child, upon finding their "real" parents, might well find that their adoptive family has far more bearing upon who they are than those who share their genetic material. While the media highlights stories of adopted children who embark upon painstaking journeys to find their "real" parents for reasons related to self-identity, a study of 715 adopted children did not rate this in the top five reasons for desiring knowledge about their parents.

11 This is consistent with Ten's interpretation. Ten suggests that in all Mill's examples, A can resume his or her life plan after preventing harm to B. "No permanent obstacles are placed to their achievement of their aims and purposes in life." This is not very demanding of agents, and I will presently suggest more demanding alternatives. My own view is that it is reasonable to intervene in A's behavior even if there is a chance that intervention will permanently frustrate A's goals, provided that that chance is small enough.

Chapter 17

A New Consultation Management Process For Managing Divergent Community Views: Lesbian And Single Women's Access To Artificial Insemination And ARTs

Leslie Cannold and Lynn Gillam

Since the birth of the world's first 'test-tube' baby in 1978, societies have grappled with the question of whether and in what ways assisted reproductive technologies should be regulated. In many countries, committees and statutory bodies have been exposed to and have often solicited through the mechanism of public consultation widely divergent views on the moral and political acceptability of ARTs. While such diversity of opinion is to be expected, and some even argue welcomed, in pluralist liberal democratic societies (Engelhardt, 1994), it is often unclear how divergent community views are and ought to be fed into the opinion-forming and decision-making processes of governments or the bodies that advise them.

This lack of clarity is evident in the way advisory bodies report on their processes. One example from Australia is the 1982 Committee to Consider the Social, Ethical and Legal Issues Arising from In Vitro Fertilization (commonly known as the 'Waller' Committee). While the Committee took submissions from 'interested parties' and experts, the report provides little information about whether and/or how such submissions informed the Committee's findings and recommendations (The Committee to Consider the Social, Ethical and Legal Issues Arising from In Vitro Fertilization, 1990, Interim Report, pp. 4-5).[1] Similarly, in the UK, organizations such as the Human Fertilisation and Embryology Authority provide little insight as to how information received from the public is used by the authority to develop policy (Haggar, 1997, pp. 12-16).[2]

In searching for detail as to the way in which advisory bodies utilized information collected during consultations, we were mindful of a problem – identified in the literature – that can plague such bodies: namely, how to deal with the radically disparate views often expressed by community members and experts who make submissions. Loane Skene, a veteran of several advisory bodies, has noted that:

> Invariably, community consultation results in numerous submissions from lobby groups, with many sending a pro forma submission devised by a leader on behalf of a group. How should the various submissions be 'weighed' and what account should be taken of them in policy development? (Skene, 1999, p. 266).

A brief glance at those submitting evidence to various consultations conducted by the Waller Committee, for example, suggests such conflicts are inevitable. Those making submission included numerous ministers of various religions, Concerned Parents, the Ethnic Communities Council, Women's Refuge Referral Service, the Humanist Society of Victoria, IVF Friends, the Knights of the Southern Cross, Pro-Life Youth, the National Catholic Rural Movement, Right to Life Victoria, the Uniting Church, the Victorian Standing Committee on Adoption, Women who Want to Be Women, The National Council for the Single Mother and her Child and the Australian Relinquishing Mothers' Society (The Committee to Consider the Social, Ethical and Legal Issues Arising from In Vitro Fertilization, 1990).

In the first section of this paper, we will discuss why we believe regulation of ARTs may be justified, and why public consultations should play a central role in the work advisory bodies undertake in making regulatory recommendations to government. In section two, we will both propose and justify a method for dealing with the contradictory moral views expressed by interested parties during the consultation process. To illustrate this method, we will use the example of the attempt by single and lesbian women to access donor insemination services and ARTs. This issue of access is contentious in many countries that offer ARTs, and is currently the subject of on-going judicial and political activity in Australia.

Justifying regulation

One response to the problem of the radical moral disagreement unleashed by public consultation is to say it indicates a good reason why the state should not seek to regulate ARTs. The view that external regulation is unwarranted has been advanced by a number of writers for different reasons. It has been argued, for example, that it is not the function of the law to enshrine or enforce particular positions on private morality. This is the case because there is disagreement within the community on such moral issues and no agreed method of establishing the moral truth of the matter, even though some people believe that there is a moral truth and that their own view represents it (Hare, 1993). This is widely taken to be the outcome of the Hart/Devlin debate over laws on homosexuality (Devlin, 1965; Hart, 1968). Indeed, it is the standard position in moral and legal philosophy is that there is 'no general direct connection between law and morality' (Lee and Morgan, 2001). However, legislation or regulation of an activity on the grounds that (some think) it is morally wrong is inadequate. Additional reasons must be provided that do not appeal to intrinsic morality. Devlin, for example, argued that the law should set only minimum standards, refrain from interfering in private matters, and only intervene where a behavior is truly intolerable, and not merely disliked by some members of the community. (Devlin, 1965). This is similar to Mill's harm principle – that we are only justified in interfering with another person's liberty when there is a genuine risk of harm to a third party, rather than merely a belief on some people's part that some particular type of conduct is morally wrong.

However, both the idea of private morality and the concept of harm are problematic. There is no clear dividing line between public and private morality, nor between conduct which does and does not harm others. What counts as a harm is not a purely factual matter, but in itself involves a value judgement. (This can be seen in the debate over female circumcision, where the loss of ability to experience sexual pleasure is seen as a significant harm by some, and as no harm at all by others.) Hence, framing public policy on the basis of avoiding regulation of matters of 'private morality' would be a fraught undertaking. Despite this we believe there is a defensible liberal presumption that the state should interfere as little as possible in the ways that citizens choose to live their lives unless a good case can be made that their conduct may cause significant harm to others. As a society, we may need to argue and reflect about what constitutes harm in a particular instance, so the liberal presumption may not always be straightforward in application. Nevertheless, this is the presumption that we believe should underpin public policy making.

On this view of law and regulation, there very definitely should not be restrictive laws about ARTs, since the degree of moral disagreement on issues related to ARTs is very high, and is tied up in a range of intimate personal matters, such as sexual relationships, reproduction and parenthood. Any moral position that was enshrined in law would undoubtedly conflict with strongly held moral convictions of some sections of the community. Further, it is not at all clear that ARTs do involve risk of harm to others significant enough to warrant interference (even if it is granted that embryos count morally as others who can be harmed). It is true that some embryos die in the ART process, but it is also true that this happens in the normal process of reproduction, and it has never been suggested that this provides a reason to intervene in those normal processes. ARTs present no more risk of harm to embryos than does sexual intercourse without contraception. The same lack of harm is evident in relation to children born as a result of ARTs. Without ARTs, these children would never have been born at all, so they can only be harmed overall by their birth if it can be shown that being born in this way was so bad that no life at all would have been preferable. It is hard to imagine how such a case could be made, except in the circumstance of very severe disability – certainly not in cases where a child is born into situation that some people think is sub-optimal (such as a single parent family).

The view about the proper function of the law will form an important part of our argument later on. However, we do not believe that it rules out any and all regulation of ARTs. Regulation of ARTs need not enshrine substantive moral views about sex, embryos, personal relationships and/or parenthood. It can be procedural, enabling rather than restrictive, and can operate in ways that represent areas of very widespread moral agreement.

Another critical response to laws about ARTs is the argument that they are an unjustified infringement on the autonomy and reproductive rights of those who wish to use these medical technologies. This seems to be Skene's point, when she makes the observation that the state does not interfere in the reproductive decisions of women who are fertile (Skene 1999, p. 266). This observation implies that the reproductive rights of the infertile are the same as those of the fertile, and questions the justification for regulating the infertile when the fertile are not similarly regulated.

We would agree that liberal democratic societies do not by and large seek to regulate the fertile (although there are some laws that do affect sexual and reproductive behavior in indirect ways, such as laws about the age of consent, and regulation of contraceptive and sterilization procedures for women with an intellectual disability). However, this is not because such regulation is, in principle, an unjustified infringement of the fertile population's autonomy and reproductive rights. Rather, it is a combination of practical and consequent moral considerations related to the different circumstances in which fertile and infertile people find themselves.

One practical consideration is the procedural difficulty of enforcing regulation of fertile people in this area. As one of us has argued elsewhere:

> The impulse to differentiate between those who should and should not be entrusted with the responsibilities of parenthood is a ubiquitous, understandable and – at times – honorable one. Who among us hasn't at some time muttered, 'there oughtta be a law' after a news report of child abuse or neglect? Thoughts of restricting the freedom to parent, however, founder on the rocks of implementation. When, how often and in what manner would the parental fitness of the enormous population of fertile people be assessed, and by whom? How would a person deemed unfit to parent be stopped from having a child: mandatory weekly pregnancy tests and forced abortions? (Cannold, 2000).

As is alluded to in the quote above, one moral consideration is privacy. Attempting to regulate the reproductive behavior of fertile people would involve a very significant on-going invasion of their privacy, in most cases out of proportion with the moral reasons which reasons that motivated the desire to regulate in the first place. So to the extent that regulation of ARTs would cause infertile people to be treated differently from their fertile counter-parts, there is good reason for this, and it does not constitute an inherent injustice: the two groups are not similar in all relevant respects, and hence there is no absolute requirement that they be treated in exactly the same way.

There are some positive reasons in favor of claims that regulation is justified. Firstly, the state has an obligation to protect the interests of its citizens and regulation is a legitimate method of achieving this. It is possible that ARTs can be practiced in ways that threaten the interests of at least some citizens and so, in principle, it is ethically permissible for the state to regulate in such situations. One example might be the use of experimental or highly risky drugs; another is use of gametes without people's consent. Both of these could occur in ART clinics, and, given the threat to the interests of individuals, the state would be justified in intervening.

Further, insofar as the state is funding infertility treatments, it has a legitimate claim (some might even say an obligation) to make decisions about how that treatment is offered. The fact that some individuals have made autonomous choices to use ARTs in particular ways does not oblige the state to give these individuals whatever they want or need. The moral power of one person's autonomy does not extend to compelling others to provide her with whatever she needs in order to fulfil her life-plan, or bring her desires to fruition. This is a well-recognized point about the limits to personal autonomy.

This is not to say that the state is justified in enacting any sort of regulation at all. One might still want to say that there are constraints, perhaps moral constraints, on the state in this regard. For example, it may be that regulation must conform to principles of non-discrimination and equality. Our point is simply that there can be justification for state regulation of ARTs – we are not yet saying anything about what the content of that regulation should be.

So we reject the idea that the difficulty faced by regulatory bodies is merely one further piece of evidence to show that ARTs should not be regulated by the state. Instead, we contend that regulation may be justified, and that while there are many different ways of regulating ARTs, advisory bodies[3] can be an excellent means of exploring and making decisions about different regulatory options. One practical advantage of this arrangement for regulation is that it provides flexibility and the ability to respond to new developments in science and medical practice in ways that legislation alone cannot. Another much more fundamental advantage is that such bodies have the capacity to play an important role in clarifying public debate on contentious matters, and contributing to broad-based and more informed decision-making by governments. Perhaps it will even lead to consensus of some sort. Consensus, of course, will not always be possible: indeed Engelhardt argues that it will never be possible in relation to ARTs (Engelhardt, 1994). This may be true, but even so, it does not mean that the process of consultation has been futile, since, as we will argue, the primary aim of consultation is to contribute to informed decision-making, not to produce unanimity in the community.

This is the role that we see regulatory bodies taking when they conduct public consultations. However, we contend that public consultation can only fulfil this sort of role if the regulatory bodies conceive of and carry out their function in particular ways. If they do not have a clear and appropriate view of their role and procedures, along the lines proposed below, they will serve merely to highlight community confusion and tension, and leave themselves without a way to sort through the morass and formulate a way forward. So we propose that advisory bodies firstly explicitly clarify what they are trying to achieve, and then set parameters for receiving and considering submissions, in terms of (a) the issues that submissions address; (b) who has a voice in the submission process; and (c) how much weight is accorded to what they say.

Clarifying the role of regulatory bodies

We suggest that there are at least three different ways in which a regulatory body could conceive of its task of public consultation. One possible option is to view the task of the regulatory body as being to advise the government of the correct moral view on the issue at hand. Thus, a body would call for submissions about the morality of access by lesbian and single women to ARTs, and would receive submissions from various groups expressing strong but conflicting moral views on the subject. The members of the body would then carefully scrutinize the content of these submissions and try to discern from them, perhaps with the aid of other sources of moral enlightenment or ethical expertise, where the moral truth lay. In such a scheme, community opinion is valued as a possible source of insight into moral truth. But this notion of being able to reliably discern the moral truth, let alone being justified in putting into legislation, has already been rejected. If lawmakers cannot justifiably claim that they know the moral truth on an issue of private morality and have no remit to put morality into legislation, then appointing a committee to find the moral truth for them would likewise be a fundamentally flawed undertaking.

A different interpretation of the task of regulatory bodies would be that it is to advise government on the state of public opinion on the issue. So regulatory bodies would in their consultation processes find out who supports and who opposes access by single women and lesbians, how strongly-held their views are, what proportion of the community holds such views, and so on. This is the sort of information that governments would find valuable, and may well use as a basis for legislation. Insofar as an advisory or regulatory body needs to provide what politicians want, then it must provide information about public opinion. But to think that this is the only aim of a public consultation is, we suggest, to have an impoverished notion of what public consultation is and what it can achieve. In terms of democratic theory, this conception of public consultation is going down the path of either majoritarianism or interest group bargaining, in which the issues are not really considered on their merits, but viewpoints are simply measured in some way, or traded off against each other. We suggest that regulatory bodies have a more robust role to fulfil than this. If they advise on public opinion, they do more as well.

We propose that the proper role of regulatory bodies is to advise government on the likely social impact of options for regulation in the area of ARTs. We will expand on the notion of social impact below, but in essence, we are referring to the effects that regulation will have on the lives of individuals and social groups. Assessing social impact requires posing and attempting to answer the question: 'How will life be for this individual or social group if ARTs are regulated in one or another way or not at all?'. More precisely, we suggest that changes to quality of life and dispositional autonomy (by which we mean the ability to live one's own life according to one's own conception of what a good life is) be used to measure social impact. Admittedly, autonomy and quality of life are moral concepts, so the consultation process which we propose would not be restricted entirely to non-moral matters. However, they are 'thin' concepts, largely without substantive content – they point to the importance of people being able to live and evaluate their own lives, but they do not prescribe any set way of doing this. Our proposal, then, is that the focus in public consultation be on how people lives will be affected, from their own point of view.

This may sound like a simple, almost trivial, idea but we believe it gives significant direction to regulatory bodies about how they should go about their work. We will demonstrate this below in relation to the issue of access to ARTs by lesbian and single women. But first we need to give an account of why social impact should be the focus of advisory bodies' consultation and recommendation processes.

Why social impact?

Our proposal that regulatory bodies should focus on the social impact of proposed regulation is based foremost on the view that those who are directly affected ought to have a say in such regulatory decisions, and that special weight ought to be accorded to their views of how they will be affected.

Having a say

Feminist theorists have probably been amongst the most influential in arguing against the legitimacy of decisions that profoundly affect women's lives but are made without women's input: decisions about abortion access, birthing arrangements and access to and quality of ARTs, for example (Cannold, 1998; Jordon, 1993; Rowland, 1987). Likewise, indigenous activists in many parts of the globe regularly assert a right to 'self-determination' – a controlling stake in decisions that affect their lives. The philosopher James Bohman takes a similar view. He argues that the goal of public deliberation is consensus, which he defines as 'the agreement of all those affected by a decision'. Whilst we do not agree that consensus is the goal, what is important here is the recognition that it is those affected who need to have a say.

Interestingly, groups which assert that those affected by a decision have a moral claim to be included in its making tend to see this assertion as self-evident. Little explanation for why this should be the case is offered. A basic understanding of the meaning and requirements of autonomy, however, seems to provide the necessary link. Young argues that an autonomous person

> ...must not be subject to external interference or control but must, rather, freely direct and govern the course of his (or her) own life. The autonomous person's capacities, beliefs and values will be identifiable as integral to him and be the source from which his actions spring...such a conception of human thought and action requires more than just the absence of constraints and instead extends to the charting of a way of life for oneself... (Young, 1986, p. 1).

An agent is 'recognised as having dignity that moral agency bestows', says Young, only if she is able to 'determine her own ends (at least to some degree)'. Moreover, from the agent's own point of view, 'autonomy promotes self-esteem' (Young, 1986, p. 2). Thus, if a person has autonomy, she has some degree of control over her life. If people have some degree of control over their life, they respect themselves and are respected by others. Be that person a pregnant woman, a citizen in a democratic state or an indigenous person, a primary path to the exercise of autonomy (and the social and personal benefits that this exercise provides) is a say in decisions that affect his or her life. People not directly affected by a decision may benefit in other ways (for instance, financially) through their exercise of control over it. But if, in exercising control, they are in fact determining what other people's life options will be, they risk using those others as a means to their own ends. This would be a stark failure in respect for autonomy, and it is vital to ensure that this does not happen. Focussing on social impact is a way of avoiding this problem.

According weight

Focussing on social impact is also important because it allows due weight to be given to those who will be most affected by any proposed regulation of ARTs.

We take the view that claims made by those most directly impacted about the effects of a certain form of regulation on them should be given extra weight, on epistemic grounds. That is, that they are in the best position to know how they will be affected, in terms of both the impact on their quality of life, and on their ability to make and carry out their own life plans.

Using the procedure: the example of lesbian and single women's access to ARTs

To demonstrate the usefulness of the procedure, we will propose a scenario in which a regulatory body conducts a public consultation on the issue of access of lesbian and single women to ARTs. We will speak hypothetically about how a public consultation might be conducted, and will not be referring specifically to any particular consultation that has actually been undertaken. In particular, we will not be referring to the consultation process that was undertaken by the Australian Senate Legal and Constitutional Legislation Committee on this issue in 2000. However, some background on the circumstances surrounding this particular consultation will be helpful in understanding our proposals, and the sorts of problems they are designed to address.

Background

Debate in Australia began in earnest regarding the permissibility of allowing single women and lesbian couples access to infertility treatments, following the 1997 ruling of the Queensland Anti-Discrimination Tribunal that allowed lesbians access to donor sperm. At the end of 1999, a little-reported Human Rights and Equal Opportunity Commission (HREOC) judgement found the Royal Women's Hospital in Victoria had discriminated on the basis of marital status when they denied a woman access to donor sperm because she was single. The Commission noted that the Commonwealth Sex Discrimination Act (SDA) prevails over discriminatory sections of State law, and awarded damages against the hospital. Not long after that, Dr John McBain brought a case in the Federal Court – in relation to his proposed treatment of a single medically infertile woman named Lisa Meldrum – and achieved a similar ruling. In McBain v. Victoria (2000), the Court ruled that parts of the Victorian Infertility Treatment Act were invalid as they discriminated against single women and were thus inconsistent with section 22 of the Commonwealth SDA.

In response to McBain, the Victorian State Government reinterpreted State law to focus on an existing provision allowing only clinically infertile couples to access ARTs. This requirement was applied to single and lesbian women, enabling only women in these categories who were medically, rather than socially, infertile to have access to the technologies.[4] This was done, said the Victorian Premier Steve Bracks, in order to preserve the State government's right to ban fertile women from seeking treatment such as donor insemination on the basis of a 'lifestyle' choice.

The Victorian opposition leader praised both the Premier's stance and the move by the Federal Government to introduce a bill into the Commonwealth Parliament to amend the Federal Sex Discrimination Act. The bill has now lapsed.[5] The amendment would have given State Governments the power to restrict access to assisted reproductive services to single and lesbian women. The Prime Minister of Australia described the problem posed by the 'McBain' ruling as 'overwhelmingly' about the 'right of children in our society to have the reasonable expectation, other things being equal, of the affection and care of both a mother and a father' (Howard, 2000). The Federal Attorney General urged the Labour Party, which opposed the bill, to 'put the interests of the children who may be born as a result of ART ahead of petty point scoring' (Williams, 2000).

The introduction of the Federal legislation, while vociferously criticized by the Federal Sex Discrimination Commissioner, was welcomed by the Australian Family Association (AFA) who praised the Prime Minister for moving 'so quickly to stop a situation which was patently against the interests of a child' (Metherell and Whelen, 2000) . In a highly unusual move, the Catholic Church and the Australian Catholic Bishops' Conference (ACBC), who were not parties to the original action, sought review of the Federal Court ruling in the High Court of Australia.[6] The ACBC described its motivation to intervene in the case as its concern for the 'welfare of the child' (Australian Catholics Bishops' Conference, 2000). The High Court granted the Human Rights and Equal Opportunity Commission and the Women's Electoral Lobby (a feminist grass-roots organization) leave to intervene in the case as the major contradictor to the Australian Catholics Bishops' Conference (ACBC), now joined by the AFA. At the last moment, the Federal Government moved to intervene in the case in partial support of the Catholic ACBC's position. The case has been heard, but a ruling is not expected until 2002.

Describing the proposed assessment procedure

The essence of the procedure is its facilitation of an advisory body focus on the impact of different regulatory options on people's lives. The critical factors for distinguishing between views and according them weight would be the directness, certainty and type of impact the proposed regulatory intervention would have on the party making the submission.

Format of submissions

The first step would be to call for submissions in such a way that social impact issues are clearly addressed. Interested parties would be encouraged to make submissions using a pro forma. The pro forma would ask the party to explicitly address they way(s) in which they or those they are seeking to represent would be impacted by the proposed regulation. The pro forma would request interested parties to articulate the nature of this impact in terms of agreed notions of dispositional autonomy and quality of life. These requests would discourage, although not prohibit, interested parties from putting forward arguments based on speculative or controversial moral claims. Of course, people making submissions cannot be forced to address the set criteria, and some may not. However, the advisory body would have a clear idea of what it was looking for in submissions as well as an explicit basis on which to interpret and weight them. Here we are not putting forward what would arguably be the more radical version of our argument, that people should be prevented from expressing moral views on issues other than those related to personal autonomy and quality of life. But we are proposing that such views be regarded as not germane to the task of the body conducting the consultation, which should be seeking to assess social impact (in the terms we have defined) and not moral outrage.

Evaluating submissions

Once submissions are received, the advisory body would use the notion of social impact to evaluate them. This evaluation would address three aspects: the directness, certainty and type of impact that is claimed in the submission.

Directness of impact Advisory bodies should first inquire about the directness or indirectness of the involvement of the party making the submission. In asking the question, 'In what way would you be impacted by the proposed regulatory intervention?', the advisory body would seek to categorize answers along a continuum.

Directly	Indirectly (1st, 2nd, 3rd Order)	Not Impacted

Figure 17.1 In what way would you be impacted by regulatory intervention?

At one end of the continuum would be those directly impacted, in the middle those indirectly impacted, and at the other end those not impacted at all. At the directly impacted end of the continuum would be individual single and lesbian women or couples, children who would be born to these women should they be given access to the technology, and the infertility treatment specialist who would provide or be prohibited from providing treatment to these women. In the middle would be those indirectly affected, who may be categorized as experiencing second or third 'steps' of impact. By this, we mean the impact on them would be contingent on the regulatory intervention first impacting on another individual or group. So for example, women as a social group might be categorized as likely to experience second degree impact. This is because the consequences they might suffer - for example, a reduction in their legal protection from discrimination on the grounds of marital status or other grounds - would only occur if the Federal Sex Discrimination Act was weakened to enable discrimination against single and lesbian women in the provision of ARTs. Third order impact might be married heterosexual couples who could argue they would be impacted if single and lesbian women were given access to ARTs and, as a second-order consequence, the legal and social status of single women and lesbian couples increased so that, as a third-order consequence, they suffered a decline in their relative legal and social relationship and parenting status. It is difficult to see how the Catholic Church, which has been highly involved in the public debate and legal activity surrounding the issue to date, could argue that it was impacted at all, or in a higher order way, by regulatory intervention around ARTs.

Certainty of impact The second category is concerned with the certainty of impact on the person or entity making the submission. In response to the question, 'How likely are you to be impacted by regulatory intervention?' the regulatory body would again seek to categorize answers along a continuum.

Very Likely	Likely	Possibly	Unlikely	Very Unlikely

Figure 17.2 How likely are you to be impacted?

Moving from the left to the right side of the continuum, there are those who will 'definitely' be impacted, those 'likely' to be impacted, those 'possibly' impacted, those 'unlikely' to be impacted and (on the far right side) those 'very unlikely' to be impacted. An advisory body might categorize single and lesbian women, children born to these women and infertility treatment specialists as 'definitely' affected. Uncertainty would likely arise about the proper placement on the continuum of other interested parties, for instance social groupings like 'women' or 'married couples' and institutions like the Catholic Church. This is because determining the likelihood of these groups experiencing predicted outcomes is not straightforward. In seeking to evaluate submissions from these groups then, the advisory body may wish to seek expert advice (see below).

Type of impact The third category is concerned with the type of impact that regulatory intervention might have. Here the advisory body would ask 'What type of impact would you experience as a consequence of regulatory intervention?'. Single and lesbian women might describe the impact regulation would have on their capacity to control their reproduction, and to fulfil their desire to have a child. Infertility treatment specialists might set out the way in which restrictive regulation would affect their professional autonomy and earning capacity, while married heterosexual couples might talk about the negative impact legislation might have on their capacity to retain sole access to the social, legal and economic benefits society now reserves for heterosexual married couples. Institutions like the Catholic Church might address the impact a regulatory intervention might have on their congregations' and the wider community's views of the sacredness of marriage, and hence the general credence given to the Catholic Church as a source of moral authority.

As we suggested above, impact would primarily be addressed in terms of dispositional autonomy and quality of life. Some of those making submissions may prefer to cast their views in terms of rights being overridden, or moral wrongs being done. For example, the Catholic church may want to argue that allowing lesbian women access to ARTs would result in children being born outside marriage, which (they would claim) is morally wrong, or infringes on the rights of children. Although we do not believe advisory bodies ought to try and prevent interested parties from expressing such views, we believe the advisory body should inform interested parties that their role prevents them from passing judgement on the truth of such claims. Further, the advisory body needs to inform interested bodies that its procedures for evaluating the social impact of regulatory options would make the systematic and transparent evaluation of such claims highly problematic.

Expert advice

The advisory group should also seek out the advice of experts on questions of fact. However, as Bayertz notes, the difficulty of reaching consensus in the area of ART is not only a consequence of pluralist values in multicultural societies, but also the ambiguity of so-called empirical facts:

> The more fragmentary the knowledge of a particular field, the greater the room for reasonable disagreement; the newer the field in question the more fragmentary the knowledge...the options are too new and their consequences too unclear for society and its members to have gathered sufficient information and to have given them sufficient moral reflection...As long as reliable answers are unavailable, there remains room for controversial estimates (Bayertz, 1994, p. 9).

Thus advisory bodies should be encouraged to be active in seeking out expert advice, and to triangulate: to seek knowledge of the 'facts' from numerous and diverse sources, so as to arrive at the most comprehensive empirical picture possible.

Submissions made on behalf of others

How should the advisory group view submissions from interested parties describing the impact of any or no regulatory intervention on other parties - perhaps ones unable to speak for themselves? In the current debate, for instance, the Women's Electoral Lobby has sought to articulate the impact on 'women' of proposed changes to the Federal Sex Discrimination Legislation, and the Catholic church has sought to describe the impact of regulatory change on children's rights and welfare. If the advisory body accepts that 'voiceless' individuals or groups like these may be impacted by proposed regulatory change, then it is imperative that the advisory body be appraised of the degree, certainty and type of that impact. But how?

We believe the advisory body should actively seek out the best individual/s or group/s to describe the impact of proposed regulation on the 'voiceless'. The best individual/s or group/s may not necessarily be those who offer themselves for the job. Asking a few simple questions, however, will enable an assessment of those who step forward. Such questions might include:

- Are you the 'closest available fit' with the individual/s or group/s being represented?
- How many members of this group do you represent?
- Have you been authorized to represent some or all of the people in this group? In what ways?
- Are you the only individual or group that can legitimately describe the impact on the voiceless?

- What is the basis of your claim to knowledge of likely impacts?

From the responses to these questions the advisory body may:

- Accept the party that stepped forward to describe the impact on the voiceless as the only party to be heard.
- Accept the party that stepped forward as only one of a range of individuals or groups that should be heard.
- Reject the party that stepped forward as unsuitable for informing the advisory body of the impact on the voiceless.

In cases where no individual or group has stepped forward to discuss the impact on the voiceless, or when the party/ies that have come forward are deemed unsuitable to solely inform the advisory body, or to advise it at all, then the advisory body ought to solicit other individuals/groups to provide it with the necessary information about impact. Individuals or groups may be solicited who are, in the advisory body's view, a good enough fit with the voiceless group that they can represent them exclusively (for instance, adult children of single and lesbian women might be solicited to discuss the way regulatory change might impact on children who might be born to single or lesbian women as a consequence of regulatory change). Alternatively, a range of 'imperfect' sources may be sought to provide as balanced a view as possible of the impact on the voiceless of regulatory intervention.

Possible objections

To some, our proposal may seem excessively restrictive of free speech: denying the right of everyone in the community to make whatever submission in whatever terms they wish and to be equally heard. However, we see this as a small loss for considerable gain, if indeed it is a loss at all. It may seem anti-democratic to restrict free speech in this way, but our view is that in a liberal democratic society, it is quite appropriate to conduct public consultation about regulation of ARTs in this way. ARTs, though they involve lots of complex technology, are about personal reproductive and family decisions, which are ultimately private matters. In a liberal democracy, the state is to refrain from interfering in the private lives of its citizens as much as possible (provided third parties are not exposed to risk of non-trivial harm). This is what liberalism is all about – allowing individuals to live according to their own conceptions of a good life. If a person makes a choice about how and when to have a child, and someone else proposes to intervene to prevent this, then it is entirely legitimate to ask 'in what way does this affect you?'. Our proposal is that every person or group wanting to make a submission should make the case. If they cannot, their claim to have their views influence the decision-making process are considerably weakened.

We would also suggest that in a consultation process where no such procedures are used to manage submissions, it is unlikely that advisory bodies give equal or appropriate weight to all submissions anyway. Rather, bodies will accord greater and lesser weight to submissions on the basis of idiosyncratic criteria that are never made public and for which they are never held accountable. Thus, while some may disagree with the specific procedures we have proposed, we believe it is incumbent upon them to propose an alternative, as the current situation where deliberations are guided by no transparent procedure is clearly inadequate. Our proposal has the merit of being straightforward, relatively easy to put in to practice, and is founded on a principled refusal to allow a single substantive moral position to dominate public policy making.

Thirdly, we note that while this proposal lacks precedent in the field of ethics, the values and objectives embodied in it have long been articulated and enacted through the legal concept of standing. In Australia, those who wish to be heard in the Courts must be able to demonstrate standing. The law of standing is the 'set of rules that determine whether a person who starts legal proceedings is a proper person to do so' (Australian Law Reform Commission, 1985, Summary). The effect of these rules is that:

> the plaintiff generally has to demonstrate some special connection with the subject matter of the proceedings – namely that he or she is specially affected, or has a special interest, or has an interest going beyond the interests of ordinary members of the public (Australian Law Reform Commission 1985, Summary).

In its 1985 report on the existing laws relating to standing, the Australian Law Reform Commission argued the law of standing should be reformed to facilitate the capacity of 'public spirited citizens' to take on the role of 'guardian of the public interest without undue restriction.' However, this requirements needed to be balanced with the 'prevention of mere meddlers' in proceedings who had 'no personal stake in the matter' and were 'clearly' unable to 'represent the public interest adequately' (Australian Law Reform Commission 1985, Recommendations).

It was on the issue of standing that numerous legal and political commentators objected to the ACBC's decision to ask the High Court to overturn McBain. Senior Lecturer at Melbourne University Kristen Walker argued that the Bishops lacked 'sufficient interest' to bring the case because the grounds on which they claimed to be personally and concretely affected by the ruling (that they provide adoption and obstetric services to married couples only) were weak. In addition, Walker questioned the capacity of the Church to adequately represent the views of 'unborn children' by noting that, '...there's very little that the Church can offer by way of any particular ability to speak for the unborn child any more than the single woman who wants to bear the unborn child can speak for that child' (The Law Report, 2001).

Standing seeks to ensure that only those whose interests are directly affected have their views considered by decision-makers. It also seeks to ensure that those seeking to represent the interests of others, especially those unable to speak for themselves, are qualified to do so. In addition, the law of standing attempts to ensure that the judicial system functions efficiently; its processes not disrupted by 'busybodies' who have 'no claim to be treated as qualified to represent in a concerned and capable manner a class or group of individuals within the community whose legal rights or whose interests, in a general sense, are directly involved in the proceeding' (Australian Law Reform Commission 1985, Summary). Whilst it is true that there has been a trend in law to expand the grounds on which standing may be granted, so as to allow the involvement of the 'public-spirited citizens' referred to the Law Reform Commission, the notion of standing itself, and the objectives which it is intended to achieve, have not changed, and are still regarded as central to legal process.

We believe that advisory bodies have similar interests in encouraging civic-minded individuals to represent the public interest on matters to do with ARTs. And likewise, such bodies need processes similar to the rules of standing, that enable them to function efficiently. They also need transparent procedures that allow due weight to be given to the testimony of those whose capacity to control their own lives may be affected by the advice they give. The procedure we have advanced in this paper for use by advisory bodies seeks to balance these requirements by enabling a differential weighting of disparate views according to the directness, certainty and type of social impact regulation can be expected to have on an individual or social group.

Assumption

One possible to response to our argument to this point is to accept its apparent legitimacy in relation to regulation of ARTs, but to question whether it would prove equally satisfactory in quite different public policy disputes, such as the debate over logging of forests in Australia, or the legality of fox hunting in the UK. If it does not seem to produce satisfactory results in those situations, it might be argued, then it cannot be a justifiable approach to public consultation and regulation in any area. This argument, of course, relies on that mechanisms for making public policy decisions must be universally applicable in order to be ethically justifiable at all. This assumption could be disputed, but we will accept it here for the sake of argument. The important thing to point out here is that our procedure, properly understood, does not produce counter-intuitive results in other areas. To take the logging example, it might seem that our procedure would give the greatest voice to timber workers who would lose their jobs if logging were to cease, and would rule environmental concerns out of court. But this is not so. One way of understanding environmental concerns is to cash them out in terms of the negative impact on future generations of environmental degradation. Hence our procedure would clearly include this concern – future generations would be affected, this does matter, and evidence would need to taken from groups or individuals that could speak to the nature and degree of impact on this voiceless group. This may not completely satisfy some deep ecologists, who attribute intrinsic value to the environment, but it is quite enough to get environmental concerns squarely on the table.

So whilst criticisms of our proposed procedure can be made, we believe that we can adequately address them. We are not claiming that this procedure is morally perfect – there are moral gains and moral losses. But we believe that the moral gains, in terms of the ethical conduct of public consultation about regulation are important, and that our procedure is the best way yet proposed of achieving these gains, with minimal moral losses.

Notes

1 The 'Waller Report' does note that 'common' IVF (where the gametes of a husband and wife are used) commands 'substantial support in the Victorian community'. It is not made clear, however, whether this support is the reason, or numbers among the reasons, for the Committee's recommendation that it be 'recognized' (The Committee to Consider the Social, Ethical and Legal Issues arising from IVF, 1990, Interim Report, pp. 27-28, 32).

2 Like the 'Waller Report', however, tantalizing clues are provided as to the meaning the Authority gives to public views. For instance, at one point it is suggested that consultation enables 'on-the-record' decision-making whereby the 'process is made as open as possible and reasons are given for policies ultimately adopted' (Haggar, 1997,

p. 12). As well, it is suggested that the consultation process enables the authority to make an 'informed decision' (Haggar, 1997, p. 13). It is not made clear, however, how the Authority understood and made use of the disparate views expressed during consultations to achieve either of these goals.

3 The remit of advisory bodies is very much determined by their terms of reference, or in the cases of statutory authorities, the description of their function set down by law. Thus their capacity to be flexible – to remain abreast of changing community attitudes or to revise their views in light of new information – can be a function of their legal status and structure. For a brief discussion of this, see (Gunning and English, 1993, p. 67).

4 Late in 2001, the Infertility Treatment Authority proposed revisions to the guidelines for eligibility for infertility treatment in Victoria. These revisions would note that 'infertility may encompass psychological as well as physical symptoms'. If enacted, such guideline revision would enable a fertility specialist to provide donor insemination to a woman who is unable to participate in heterosexual intercourse. This would include women who cannot have sex with a man; partners in a lesbian relationship; women in a heterosexual relationship (whether married or not) and; single women who have been unable to sustain a heterosexual relationship because of this condition. 'Political concern about the proposal led the ITA to withdraw the suggested revisions and the State Liberal opposition to promise legislation that would restrict ART access to those who are physically infertile' (Infertility Treatment Authority, 2001, p. 58; McNair, 2001, p. 57; McNair, 2001, p. 53).

5 The Bill lapsed because it was not passed by the Senate prior to the dissolution of Parliament for the November 2001 election. Walker (2001) argues that the re-election of the Liberal/National Party Coalition makes it likely the Bill will be reintroduced, but the balance of power in the Senate means it remains unlikely to pass. If it were to pass, it would revive the operation of the Victorian legislation and thus reinstate the exclusion of lesbians and single women from access to assisted reproductive services (For further discussion of the legal issues involved, see Walker, 2001, p. 56.)

6 Walker notes that it was not possible for the Church or the Bishops to appeal as they were not parties to the proceedings, having elected only to present arguments as amici curiae. In seeking review, they utilized procedures under s 75(v) of the Constitution and s 30 of the Judiciary Act 1901 to attempt to have Sundberg J's decision quashed by the High Court. She believes there is 'some doubt about whether the Court will permit the ordinary appeals process to be circumvented in this way'. As well, doubt exists as to the standing of the Bishops and the Church to seek review. However, she notes that 'the need to demonstrate standing was avoided in part by the grant of a fiat to the Church by the Commonwealth Attorney-General. The fiat allowed the Church to proceed with the case in the Attorney's name without needing to demonstrate standing' (Walker, 2001, p. 56).

References

ABC Radio, (2001), 'Catholic Bishops and IVF: Teaching Western Culture', *The Law Report*, Radio Program.

ABC TV, (2001), Family Matters - Birthrights. *Sydney: Compass*. Television Program.

Australian Catholics Bishops' Conference (2000), 'Catholic Bishops will urge Court to uphold existing laws supporting essential bond between parents and children'.

Australian Catholics Bishops' Conference (2000), 'McBain Ruling raises fundamental

questions regarding family, marriage and the rights of the child'.
Australian Law Reform Commission, (1985), *ALRC 27 - Standing in Public Interest Litigation*, National: Australian Law Reform Commission.
Bayertz, K. (1994), 'Introduction' in *The Concept of Moral Consensus*, K. Bayertz (ed.), Kluwer Academic Publishers, Dordrecht, Boston, London, pp. 1-18.
Bohman, J. (1998), 'Survey Article: The coming of age of deliberative democracy', *The Journal of Political Philosophy*, Vol. 6, No. 4, pp. 400-425.
Cannold, L. (1998) *The Abortion Myth: Feminism, morality and the hard choices women make*, Allen & Unwin, NSW.
Cannold, L. (2000), 'Calling All Single Childless Women', *The Age*, 20 March.
Devlin, P. (1965), *The Enforcement of Morals*, Oxford University Press.
Dutton, Re. A. (2001), Personal Communication.
Engelhardt, T.H. Jr., (1994), 'Consensus: How much can we hope for? A conceptual exploration illustrated by recent debates regarding the use of human reproductive technologies', in *The Concept of Consensus: The case of Technological Interventions in Human Reproduction*, K. Bayertz (ed), Kluwer Academic Publishers, Boston and London.
Gough, K., Stephen B. and Marris S. (2000), 'State doctors to get IVF legal advice', *The Australian*, 1 August, p. 6.
Gunning, J. and English, V. (1993), *Human In Vitro Fertilization: A case study in the regulation of medical innovation*, Dartmouth Publishing Company, Aldershot.
Gutman, A. and Thompson D. (1990), 'Moral Conflict and Political Consensus', *Ethics*, Vol. 101.
Gutman, A. and Thompson D. (1995), 'Moral disagreement in a democracy', *Social Philosophy and Policy*, Vol. 12, No. 1, p. 106.
Haggar, L. (1997), 'The Role of the Human Fertilisation and Embryology Authority', *Medical Law International*, Vol. 3, pp. 1-22.
Hart, H.L.A., (1968), *Law, liberty and morality*, Oxford University Press, Oxford.
Howard, J. (2000), *Sex Discrimination Act, IVF and States' rights*, Parliament House, Canberra.
Jordon, B. (1993), *Birth in Four Cultures: A cross cultural investigation of childbirth in Yucatan, Holland, Sweden, and the United States* (4th edn), Waveland, Prospect Heights, Illinois.
Lee, R. and Morgan, D. (2001), *Human Fertilisation and Embryology: Regulating the Reproductive Revolution*, Chapter 2, Blackstone, London.
McClelland, R., Macklin, J. and Laurence, C. (2001), 'IVF Amendments are Bad Law'.
McNair, R. (2001), Personal Communication.
Metherell, M. and Whelen, J. (2000), 'PM Fights IVF for Singles', *Sydney Morning Herald*, 2 August, p. 1.
Rowland, R. (1987), 'Making Women Visible in the Embryo Experimentation Debate', *Bioethics*, Vol. 1, No. 2.
Skene, L. (1999), 'Why Legislate on Assisted Reproduction?' in *Controversies in Health Law*, I. Freckelton and K. Peterson (eds.), Federation Press, Melbourne.
South Australian Council on Reproductive Technology (2000), *Quarterly Bulletin*, South Australian Council on Reproductive Technology, Adelaide.
South Australian Council on Reproductive Technology (2001), *Quarterly Bulletin*, South Australian Council on Reproductive Technology, Adelaide.
The Committee to Consider the Social, Ethical and Legal Issues Arising from In Vitro Fertilization (1990), *Consolidated Reports of the Victorian Inquiry into IVF and Related Issues 1982-1984*, The Committee to Consider the Social, Ethical and Legal Issues Arising from In Vitro Fertilization, Melbourne.

Williams, Attorney General D. (2000), 'Sex Disrimination Amendments Pass the House'.
Young, R. (1986), *Personal Autonomy: Beyond negative and positive liberty*, Croom Helm, London, Sydney.

Chapter 18

The Australian Experience Of Self-Accreditation

Douglas Saunders

Introduction

Australia was very early in the acceptance of a need for the new Assisted Reproductive Technology (ART) to have some kind of overview, including peer review. We now have a very effective peer-review mechanism in place, called the Reproductive Technology Accreditation Committee (RTAC), which is a sub-committee of the Fertility Society of Australia (FSA). The history of how this came about, and how it gained medical, governmental and community acceptance without the need to set up some Government Authority, makes interesting reading, and, before the history is lost from memory, a brief review might be considered appropriate at this time.

Some idea about how the FSA itself was established as a scientific society in 1982 should also be considered in this context. The Society evolved from a relationship between the Australian Obstetrical & Gynaecological Research Society, the Australian Society for Reproductive Biology, and the National AID Workshops. The details were reviewed previously by Johnston, Jones and Saunders, 1991. Each society brought different approaches to the new Society.

The Australian O&G Research Society, at that time, consisted mainly of university academics wishing to conduct and promote research and, at the time of Louise Brown's birth, this heavily favored reproductive endocrinology and infertility. There were prizes presented at annual scientific meetings, and free communications were published. The Australian Society for Reproductive Biology brought scientists and the scientific advances in in vitro fertilization (IVF) that were now being rapidly applied to human infertility. The National AID Workshops incorporated a multi-disciplinary approach to the clinical problem, and established a forum for the interchange of ideas between gynaecologists, andrologists, urologists, and micro-surgeons with nurses, counselors and lawyers.

There was some initial resistance to the formation of "another Society", but the rapid developments now taking place made a combined approach inevitable. The FSA then included a consumer representative (a new organization named ACCESS) on its Council, and its aims were soon recognized by the NH&MRC, State and Commonwealth Governments, and by the Royal Australian College of Obstetricians & Gynaecologists.

The first Scientific Meeting of the Fertility Society of Australia was in Sydney in 1982, and incorporated all of these elements. It was modeled on the American Fertility Society (now renamed The American Society for Reproductive Medicine, ASRM) and had as one of its aims "to improve the standard of clinical care to patients".

One of the initial tasks of the FSA was to try and develop program standards for IVF Units, which were then proliferating throughout the country. These standards developed from the New South Wales Department of Health's requirements, and the 19 points were first published in the FSA Newsletter in 1985. Now that those minimal standards had been set, a way of peer-review to ensure that the standards were complied with had to be developed.

In 1987 RTAC was established under the Chairmanship of Professor Lloyd Cox, of Adelaide. The initial Guidelines (now 2 pages) had a requirement that all programs be subject to scientific and medical audit by RTAC. A 3-year inspection, audit and accreditation process began which has now continued under the subsequent Chairmanships of Bryan Hudson, Warren Jones, Gordon Baker and Ian Johnston. I have been the current Chair since 2001.

Composition of RTAC

The Chair and Deputy Chair are medical infertility specialists and are appointed by Council of the FSA for 3 years (with re-appointment for a further 3 years if appropriate). The organization representing the reproductive Scientists (SIRT) the reproductive Nurses (FNA), and the Counselors (ANZICA) also elect members to RTAC for a similar duration. To avoid possible conflicts of interest, the inspecting party usually does not include a representative from within the same state. The Royal Australian & New Zealand College of Obstetricians & Gynaecologists (RANZCOG) also appoints a representative, who has the additional responsibility to overview the relationship of the Unit under inspection with the other

College Fellows, and also with neonatologists, looking in particular at IVF complications and obstetric impact. In those Units with a College Trainee, (CREI Fellow or Registrar), their experience may be reviewed during an inspection.

Guidelines

The first mention of guidelines was published by the NH&MRC as Supplementary note 4 in October 1982. It ensured that all centers offering an IVF program "should have all aspects of the program approved by an institutional ethics committee (IEC)".

The program standards, which were adopted by the FSA in 1986, referred to this Document and the IEC. When the first "Guidelines to the Code of Practice for Units using in vitro Fertilisation and related Reproductive Technologies" were circulated in 1988, it referred to the then established RTAC (from 1987) and a series of explanatory notes. In 1992 Terms of Reference for RTAC were established. These were revised in 1997 and again in 2001.

The Terms of Reference are:

1. To be responsible for the Code of Practice for Centres using Assisted Reproductive Technology.
2. To review applications and re-applications for accreditation and to visit Centres to assess compliance with the Code. To grant accreditation with or without conditions for a period not exceeding three years.
3. To set standards for Clinics and to monitor achievement of such standards. Where appropriate to provide feedback to the ART Centres where practices may be enhanced. To encourage adherence to best practice principles (Quality Management Systems).
4. To publish
   - The Code of Practice for Centres using ART
   - An Annual Report of activities
   - Recommendations for alterations of the Code of Practice.

The document now is about 80 pages in length and includes 19 attachments.

Inspection process

Units are usually accredited for 3 years, but have been on occasions accredited for only one year, with a re-inspection at that time. Accredited Units have access to federally subsidized pharmaceuticals so when a new Unit is set-up a provisional accreditation is given for 12 months to allow the Unit to obtain free gonadotrophins and to prepare for formal accreditation. About 6 weeks before the visit, the Medical Director receives a Questionnaire. When completed, the details are sent to the RTAC inspecting party. The questionnaire also refers to the previous inspection, and asks how any previous problems were addressed.

The pregnancy results are also included for each ART treatment; A Gold Standard sub-group (women aged 25-34 with no previous baby from ART having their first or second stimulated cycles only) is also reported concentrating on Live Birth pregnancies per 100 oocyte retrieval cycles. Multiple Pregnancy rates are also reported, as well as admissions to Hospital with Ovarian Hyper-stimulation Syndrome. Small Units can combine two years to get significant numbers. RTAC has a policy of recommending 1-2 embryo(s) only transferred, with a 20% twin rate.

The inspection party usually has two gynaecologists, one scientist, one nurse, one counselor and a consumer representative. Each follows a prescribed inspection checklist and inspects their relevant areas. The FSA has a formal agreement with the Unit to allow this process to occur and visiting inspectors sign a confidentiality agreement with the clinic. A group of consumers is also welcomed and interviewed in the absence of the Unit staff. This inspecting process is within the requirements of the *Commonwealth Privacy Act* of December 2001.

After the inspection, which may include Satellite and Transport Units, an informal feedback session is conducted, but a consensus formal report is drafted and sent to the Unit after 6 weeks. In the states with statutory authorities set up under State legislation (Victoria, South Australia and West Australia), representatives are allowed to visit with RTAC inspectors, and to receive the final report. This saves the Unit having a double inspection.

Units fall into four categories:

1. Primary: Unit provides all routine clinical services (may or may not include pre-implantation pre-natal diagnosis).
2. Associate: Associated with a Primary Unit, for Quality Control, Staff Education and Ethics Support.

3. Satellite: Unit having an irregular operation, with staff etc supplied from the Primary Unit.
4. Transport: Medical Center, with no embryology on site.

Problems frequently arise where there has been a re-alignment of Units and their satellites. The list of Accredited Units is available for public inspection via the RTAC page on the FSA web-site at http://www.fsa.au.com.

National Perinatal Statistics Unit

From the earliest Standards and Codes the importance of data collection was recognized. The National Perinatal Statistics Unit (NPSU) which was initially set up in association with Professor Paul Lancaster has collected this data from the first Australian IVF pregnancy, in a unique data set. The NPSU publishes yearly reports.

These data will soon be available to RTAC at six monthly intervals in electronic format, enabling a more specific assessment of each Unit's Questionnaire response. The public accountability of the Units via these Reports has encouraged public acceptance of ART and this has led to public funding of much of IVF including the drugs.

Advertising Code of Practice

Also available on the FSA web-site is the Advertising Code of Practice. This particularly refers to a Unit's consumer information and web-site. RTAC endeavors to ensure that a Unit complies with this Code, and does not quote comparative "success" rates which may mislead the patients. Patients need to be informed of the dangers of statistics, particularly when the numbers of patients in the study is small, or massaged to look better. League tables as published in USA or UK are not regarded by RTAC as being in the patient's interest. Whilst it is recognized that a Unit cannot always stop "good news" stories and "breakthroughs", scientific success is best published first in a peer-reviewed journal or abstract.

Funding

RTAC inspectors work in their own time on a voluntary basis (except for the consumer representatives who are paid a small per-diem allowance).

The IVF Directors provide a budget to pay for the travel costs of the inspecting party, according to a formula derived from whether or not the Unit is a Primary, Associate, Satellite or Transport Unit. Costs vary with the distances to be traveled (including New Zealand), so some years are more costly than others. Several approaches have been made to the Commonwealth Government for a grant, but no money has ever been forthcoming. Therefore Government has had no input into the control or running of RTAC, a state which has had its benefits when it comes to independence.

Conclusion

What are the strengths of RTAC in 2002? Since 1987 when RTAC was established as a sub-committee of FSA, but funded by the IVF Directors, it has demonstrated that it can produce effective peer review. Such has been the credibility that for mainstream infertility services, there is minimal public concern. Each year the number of births after assisted conception increases. In Australia, in 1999 (the last NPSU Report available) 4,319 births occurred after assisted conception, accounting for 1.7% of all births.

Accreditation by RTAC is mandatory in those states with Statutory Authorities. Accredited Units receive free gonadotrophins (for Australian Medicare cardholders), and IVF procedures are, in part, refundable by the Australian Health Insurance Commission and private health funds (for hospital and operating room costs). RTAC has also demonstrated a capacity to respond to new ideas and suggestions, and aims to constantly improve the inspection and accreditation process.

References

Hurst T., and Lancaster P. (2001), *Assisted Conception. Australia and New Zealand, 1999 and 2000*, AIHW National Perinatal Statistics Unit, Sydney cat. no. PER 18.

Johnston I., Jones W. and Saunders D. (1991), 'The Embryo Comes of Age: The First Decade of the Fertility Society of Australia', *Aust NZ J Obstet Gynaecol*, Vol. 31, pp. 55-57.

NHMRC (1982), *NHMRC Statement on Human Embryo Experimentation*, AGPS, Canberra.

Chapter 19

Can The Use Of Human Embryos For Research Be Banned Or Would That Be Unconstitutional In Australia?

Mark W. Richardson

Introduction

In Australia, there has been recent debate and inquiry into the scientific, ethical and regulatory aspects of human cloning.[1] Had they been implemented, recommendations by an all-party Parliamentary Committee would have resulted in Australian scientists being regulated as to the manner in which they would be able to use spare IVF embryos to create new stem-cell lines. There would have been a ban on cloning for reproductive purposes but not otherwise. However, the Federal Cabinet has set Australia on a course of banning all such embryo use and research. The proposed ban would be intended to prohibit "destructive" research into human embryo stem cells. It is not clear how any such ban could be implemented.[2] The Australian Constitution is the framework within which the decision making process, which has been commenced, must be conducted and to which it must conform.

The rule of law does not mean that we must be ruled by laws. It does not mean that each and every community issue can be resolved by the passage of legislation. The starting point is, and must remain, that use of spare embryos for research is not illegal in Australia unless and until a law makes it so. It is safe to say that the law in Australia is not consistent between the various States and Territories It is for the proponents of regulation to make the case for banning, not for others to make the case for continued use.

The Church, as well as groups like Right to Life and members of the public, have made various submissions to the inquiries. Although they attempt to put their objections on a basis which is pluralistic in nature, they inevitably need at some stage of their argument to concentrate on the definition of humanity. To say that a moment after conception the entity has the essence of humanity is to adopt a religious perspective. The use of spare embryos and the associated stem-cell debate may be a fresh test of the Church's capacity to participate in a scientific discourse without the "tragic reciprocal misunderstanding" of the Galileo case.[3] As will be seen, in Australia the Church, and others from a clear religious perspective, have been keen to stress the "common morality" and that their objections to the use of embryos are not merely based on the Bible or religious convictions.[4]

Given the inherent likelihood of a religious context for the debate, the query is whether section 116 of the Australian Constitution will have any impact on how prospective legislation would be drafted. That section is described in its sidenote as "Commonwealth not to legislate in respect of religion". The question posed, would a ban on embryo research amount to legislation contrary to a constitutional freedom of religion?, is therefore topical and fundamental to Australia. In proposing an answer for Australia, issues applicable to other jurisdictions are traversed.

There will be first a brief discussion of the notion of personhood, then a reflection as to whether any one could call on a Court with relevant power to exercise its parens patriae jurisdiction in favor of a research embryo. The focus will then shift to the constitutional question derived from the terms of section 116 of the Commonwealth of Australia Constitution.

Personhood

The definition of personhood is difficult, if not problematical. Use of the term may avoid the religious overtones associated with soul or spirit which it is said differentiates humans from other mammals.[5] There is presently a tremendous sense of artificiality about a factor which is meant to be the defining essence of being human. Defining personhood is not any easier with advances in medical skills, techniques and technology. A common factor in every definition is seemingly an element of arbitrary cut off. The cut off is adopted which best suits the particular proponent's argument.

If personhood is from birth how is the different status justified? Why is a baby born 10 days early as a breach baby immediately a person; yet had he remained in utero for the expected duration of the pregnancy then during those extra 10 days in utero he would not have been clothed in the mantle of personhood? Similarly if a baby was born 7 days "late" - she is effectively denied the label of personhood for those extra 7 days.

The use of the suffix '-hood' is usually meant to mean the entity has the status of the particular noun (in this case person) or that it has the quality of being, in this case, a 'person'. To suggest that the moment after conception the cluster of cells either has the status of being human, or even has the quality of being human, simply begs the question which use of the term is religiously trying to avoid.

Indeed, it is necessary to consider whether the concept of personhood, in contrast to an entity with a legal personality attracting legal rights, is even known at law. It should not surprise anyone that much of the case law on the question of the status of the unborn arises in three fields of legal inquiry. First, succession; second, criminal; third, abortion.

Succession

The question of the status of the unborn at succession law has been settled since at least 1724 where it was said by Lord Hardwicke "The plaintiff was *en ventre sa mere* at the time of her brother's death, and consequently a person *in rerum natura* so that both by the rules of the common law and civil law she was to all intents and purposes a child".[6] So for three hundred years it has been the common law of England that the property, real and personal, of a child *en ventre sa mere* will be protected. As relatively recently as 1907 the House of Lords had to consider the meaning of "born in my lifetime" as used in a will. Lord Loreburn LC accepted as "certain that a child *en ventre sa mere* is protected by the law, and may even be a party to an action".[7] However it was held that the rule of construction by which a child *en ventre sa mere* is considered to be born or living is only valid if such a construction was necessary for the benefit of the unborn child and to enable it to take a benefit to which, if born, it would be entitled. It appears this principle is to protect a child in the womb at 3 weeks gestation as much as a child in the womb at 39 weeks gestation. The real and personal property may be protected but what of the child *en ventre sa mere*?

Criminal

Before turning to the question of abortion it has been since Coke's time that an unborn child received protection for injuries done to him in utero if he survived his birth. Murder, at common law, required the person killed to be a "reasonable creature in being and under the King's peace". Blackstone in his Commentaries amplified this by saying[8] "To kill a child in its mother's womb is now no murder but a great misprison; but if the child be born alive and dieth by reason of the potion or bruises it received in the womb, it seems, by the better opinion, to be murder in such as administered or gave them".

Crystallizing of rights The law avoids discussion of personhood or the nature of the entity in utero by defining a legal personality for that entity, irrespective of its nature. The criminal law is describing an entity which, in having a legal personality, can acquire greater (or at least enforceable) rights after an event (being born alive) than it had before such an event. This concept of the unborn having rights which come, if you like, to be crystallized at birth was discussed, and largely accepted, in the developing law of negligence in Australia. In Victoria the principle case establishing the proposition is Watt v Rama[9] - a case involving injury (as a consequence of a car accident) to the child whilst in utero and who was subsequently born alive yet damaged. The majority rely upon the birth of the child as the factor which crystallizes the duty of care at tort and say:[10]

> But as the child could not in the very nature of things acquire rights correlative to a duty until it became by birth a living person, and as it was not until then that it could sustain injuries as a living person, it was, we think, at that stage that the duty arising out of the relationship was attached to the defendant, and it was at that stage that the defendant was ... in breach of the duty to take reasonable care to avoid injury to the child.

In a separate judgment Gillard J placed initial emphasis on the need, in the law of tort, for concurrence of the injury and damage and that no liability existed until the damage was done. His Honour, like the majority, primarily relies upon the cause of action not being complete until damage has been suffered (which is not until the child is born) to found the basis for the plaintiff's entitlement to damages. He does discuss in some detail the "fiction" of rights being conferred on a child *en ventre sa mere*. Although not finding it necessary to determine whether the fiction should be applied his Honour appears to be supportive of the concept when he says:[11]

> From these cases it must be accepted that there is a rule of law which recognizes that an unborn child may possess rights. This implies that there are correlative duties imposed on others in favor of the unborn child. It also implies that an unborn child is deemed to be a "person" where the right is for his benefit and obtains the requisite capacity to institute proceedings for an infringement of his right, he may do so and obtain a remedy for the infringement when he was *en ventre sa mere*.

Abortion

With at least a recognition that the unborn may possess rights (although only if it reaches a capacity - through birth - are the rights of any real value) the issue turns to personhood and abortion. This is a topic which is far too vast to cover in the present discussion. I will only refer to three cases relevant to the Australian position - one English and two Australian cases. They each deal with applications for injunctive relief by the father to prevent the mother from aborting their child.

In Paton v British Pregnancy Advisory Service,[12] the husband claimed a right to have a say in the destiny of his child and sought an injunction to prevent his wife undergoing a lawful abortion. The Court says, of the right of the fetus that it

> cannot, in English law ... have any right of its own at least until it is born and has a separate existence from the mother. That permeates the whole of the civil law of this country (I except the criminal law, which is now irrelevant), and is, indeed, the basis of the decisions in those countries where law is founded on the common law, that is to say, in America, Canada, Australia ...

Having concluded that the fetus had no right to bring an action, the Court then determined the husband had no common law or statutory right to have a say in the destiny of the child he helped conceive. The Court did not consider the anomaly of the law granting rights to the fetus by way of succession, for example, but the law not being prepared to protect that fetus so that it could later enjoy those rights.

The same argument (husband seeking to prevent his wife having an abortion) came before Gibbs CJ, as a single judge of the High Court of Australia in 1983.[13] Gibbs CJ, in this case of *Kerr*, and "as present advised" agreed with the judgment in *Paton*.[14] There is not a flavor from the judgment that there was a great deal of intellectual debate about the issue. The husband also sought to invoke the parens patriae jurisdiction of the Court and that aspect is returned to below.

Finally, in a case before the Family Court of Australia in 1989,[15] the husband put his case on a number of bases at common law including his "right to procreate"; alternatively the right of the unborn The Court simply relied upon the judgments in *Paton* and *Kerr*. The husband then sought to argue that the Family Law Act gave power to the Court to protect incipient and inchoate rights.[16] The Court said the right being sought to be protected was "the right to be allowed to come into existence, which right I have held does not exist".[17] Later the Court said "as a fetus has no legal personality it is, in law, a non-person, and it therefore cannot have an inchoate legal right any more than it can have a fully developed one".[18]

Such comments do not sit easily with the comments made by Gillard J in *Watt*. There is a fundamental difference between the view that an entity can have legal rights which can be enforced at a later stage (from which it can be argued the entity must have had a legal personality at the time of acquiring those rights) and the view that an entity is a non-person and has no rights.

The discussion about fetal rights is difficult enough when it is conducted in the context of the mother carrying the baby in utero.[19] The complexity is heightened when you try and address the question of rights for something which is the equivalent of what is produced by natural conception but is never implanted into a womb. It has been suggested that one of the reasons why fetal rights have not been given precedence is because of the imposition upon the mother of enforcing those fetal rights. If it is accepted that the embryo/fetus does have rights (but those rights rank effectively behind the rights of the mother) then when the embryo is developed out of utero, or it is otherwise spare, then the burden of the mother's rights is not present. What is to prevent the fetal rights now being given full expression? One way would be to say that a fetus developed out of utero or otherwise spare attracts different (and presumably lesser) rights. This would indeed be a very artificial distinction. Could a Court protect any rights of an embryo created in a laboratory where there was never any intention of allowing the embryo to develop to gestation or where it was otherwise spare?

Parens patriae

This brings us to a consideration of the parens patriae jurisdiction found in at least the Supreme Courts and the High Court of Australia. The jurisdiction was originally a prerogative of the Crown to care for persons of unsound mind. However, it is an "inherent jurisdiction to do what is for the benefit of the incompetent. Its limits (or scope) have not, and cannot, be defined".[20] The breadth of the jurisdiction has been widely accepted and no list detailing the categories of cases to which the jurisdiction could attach is exhaustive. As has been said, "(that) the courts are available to protect children from injury whenever they properly can is no modern development".[21] A leading case in Australia, with respect to the parens patriae jurisdiction and sterilization, is the High Court's decision in Marion's case.[22]

There have been recent cases dealing with the jurisdiction in contexts outside seeking approval for sterilization of incompetent individuals. In Northridge v Central Area Health Service[23] the hospital gave, over the objection of the family, non-resuscitation orders. In discussing the parens patriae jurisdiction O'Keefe J in the New South Wales Supreme Court accepted that it was essentially protective in nature and gave to the Crown "not only the power, but also the duty, to protect the person and property of those who are unable to do so for themselves".[24] His Honour adopts with approval a passage by Brennan J in Marion's case[25] in support of the proposition that "the court is empowered to protect the human dignity and rights of individuals who are disabled in such away that they cannot protect such dignity and rights for themselves".[26]

Six months earlier O'Keefe J had had to consider the application of the parens patriae jurisdiction in the context of a wife seeking to take sperm from her comatose husband.[27] The application was refused. His Honour again stressed the protective nature of the jurisdiction and that the "paramount consideration is the promotion of the health or welfare of the subject of the exercise of the jurisdiction".[28] His Honour considered that the request for semen removal was a non-therapeutic surgical procedure which would not preserve the patient's life nor which would "safeguard, secure or promote or prevent the deterioration in, the physical or mental health of the patient".[29] The best interests of the prospective child were considered and his Honour was unable to conclude that "a child's best interests would be served by being brought into existence in the manner, at the time and in the circumstances contemplated as possible by the plaintiff".[30]

It is to be noted that although the best interests of the prospective child were considered "and should under our law, be had regard to"[31] the parens patriae jurisdiction was not being exercised on behalf of that prospective child. In the context of using embryos for research we come back to the question of whether the embryo has any rights. If it does, but not until it is born, the question then becomes whether an embryo has sufficient legal personality to become the subject of the parens patriae jurisdiction. The cases speak of individuals, children and adults. And yet if the embryo does have rights which do not crystallize until birth, what greater being of disability, unable to look after its own future interests, could be imagined? Rights but no one capable of ensuring that such rights are able to be of benefit to the entity. If that is not sufficient, consider the prospective use of the spare or created embryo - research to obtain and understand stem cells with the certain and associated destruction of the embryo. In the context of its creation (being either deliberately made for research or alternatively surplus to reproductive requirements) what procedure, other than destructive research (which does not at all equate with useless research), could be less well designed to promote the welfare of the entity upon which the procedure is being performed?

As noted previously, in *Kerr* the husband sought to invoke the parens patriae jurisdiction. Gibbs CJ dealt with it on the basis that the fetus had no right of its own until it was born (accepting the decision in *Paton*). His Honour did go on to say that even if that view was wrong he would not grant an injunction saying "There are limits to the extent to which the law should intrude upon personal liberty and personal privacy in the pursuit of moral and religious aims".[32] In contrast, the Judge in *Paton* said "In the discussion of human affairs and especially abortion, controversy can rage over the moral rights, duties, interests, standards and religious views of the parties. Moral values are in issue. I am, in fact concerned with none of these matters. I am concerned, and only concerned, with the law of England as it applies to this claim".[33]

The embryo research debate is based on individual moral and ethical perspectives of each proponent and such perspectives are influenced by the person's religious ideals (whether such beliefs are based on theistic or non-theistic theories, belief or non-belief in religion). Which brings us to section 116 of the Constitution of the Commonwealth of Australia.

Section 116 of the Constitution

Section 116 is a constitutional provision of high importance.[34] It deals with the making of laws and not the administration of laws.[35] It has been labeled a constitutional guarantee, albeit limited.[36]

In its entirety section 116 is as follows:

> The Commonwealth shall not make any law for establishing any religion, or for imposing any religious observance, or for prohibiting the free exercise of any religion, and no religious test shall be required as a qualification for any office or public trust under the Commonwealth.

Its history is interesting. Originally, at the convention debates in the 1890s the clause was directed to the States of Australia - "A state shall not make any law prohibiting the free exercise of any religion". Eventually, after much debate, where the issue of religious observance seemed to be solely linked to what was done on Sundays, the Clause as presently drafted was approved. As Mr Higgins said, it was his intention

> to make it clear beyond doubt that the powers which the states individually have of making such laws as they like with regard to religion shall remain undisturbed and unbroken, and to make it clear that in framing this Constitution there is no intention whatever to give to the Federal Parliament the power to interfere in these matters.[37]

Given such intention it is curious to describe the section as a constitutional guarantee of freedom of religion, its exercise and observance. It is a strange guarantee of freedom which can be so easily overcome, in any shape or form they chose, by the legislatures of each and every one of the States comprising the Commonwealth of Australia. It is worth bearing in mind that the scope of the legislative powers of the Victorian Parliament, for example, are wide ranging and in the following terms "The parliament shall have power to make laws in and for Victoria in all cases whatsoever".[38]

Section 116 has been considered by the Australia's High Court on four occasions - in 1912,[39] in 1943,[40] in 1980[41] and recently in 1997.[42]

The 1912 case was brought by a senior cadet who argued that military drills were contrary to his religious beliefs and therefore affected his ability to have free exercise of his religion. Even in the knowledge that High Court judgments were at that time often noted for their brevity, the comment by Barton J that "I think this objection is as thin as anything of the kind that has come before us"[43] still carries the ring of the short sharp judicial slap which leads us to pass on to the next case.

The 1943 case dealt with wartime regulations affecting the Jehovah Witnesses and their advocacy of doctrines which, though advocated in pursuance of religious convictions, were prejudicial to the prosecution of the war. Latham CJ was more expansive in his consideration of the section. His Honour noted the section is a general prohibition applying to all laws made by the Commonwealth and it "prevails over and limits all provisions which give power to make laws". Of importance for the present argument (involving as it does the age old debate between religion and science) Latham CJ was of the opinion that the section operated not only to protect the freedom of religion "but also to protect the right of a man to have no religion .... proclaims not only the principle of toleration of all religions, but also the toleration of absence of religion".[44]

In this context Latham CJ stressed the section protected not just the exercise of religion but the *free* exercise of religion.[45] It must follow (just as freedom of religion, according to Latham CJ, had as its corollary the freedom to have no religion) that the free exercise of religion has as its corollary the free exercise of no religion. His Honour discussed the line of reasoning from cases in the United States, particularly those before 1900, which had effectively come to the conclusion that the equivalent provision in the United States' Constitution was not an absolute right for free exercise of religion. The United States provision "did not involve a dispensation from obedience to a general law of the land which was not directed against religion". Rather it had to be determined whether the law in question had "unduly" infringed that freedom of religion. His Honour accepted the rationale of the American line of reasoning and stressed that it was for the Court, and not for the Parliament, to decide whether the law was appropriate. Ultimately, however, his Honour determined the case on a narrower basis.[46]

Section 116 next came before the High Court in 1980 in the case dealing with Commonwealth funding of non-government schools. The phrase considered in this case was "for establishing any religion". The critical word is "for" which also begins the phrases dealing with religious observance and free exercise of religion. According to Barwick CJ the use of the word "for" meant "the law to satisfy the description must have that object as its express and, as I think, single purpose".[47] Other members of the majority[48] all considered the use of the word "for" meant it was necessary to look at the purpose of the law and purpose was not satisfied by the "mere circumstance that the law is one which touches or relates to the subject matter".[49] Only Murphy J equated the use of "for establishing" with the words in the United States Constitution "respecting an establishment of" and so gave the word "for" a broader meaning than the majority.

The "high importance" of the constitutional provision tends to lose some gloss if the prohibition contained within it can be overcome not only by individual States but by the Commonwealth itself by ensuring the single purpose of any legislation does not contravene the prohibition. It may be understandable how an Act for establishing religion could be drafted to only have that as its single purpose. It is easier, however, to envisage situations where legislation may not be for the single purpose of affecting religious freedom or imposing religious observance and yet may still achieve such ends.

The construction of section 116 and in particular the word "for" was considered by the High Court in 1997 in considering the constitutional validity of a Northern Territory Ordinance for removal of aboriginal and half-caste children to reserves and institutions.[50] Brennan CJ was of the view that section 116 could only be invoked if the law had the purpose of achieving an object forbidden by section 116.[51]

Dawson J,[52] agreeing with Gummow J,[53] also said "for" meant it was necessary to look at purpose, as the end to be achieved. Purpose was not underlying motive but the end or object served. Toohey J was prepared to accept that legislation may have more than one purpose. In the opinion of Gaudron J the clause dealing with free exercise of religion would be contravened by a law which either banned the free exercise of religion or prevented it.[54] Her Honour also accepted "purpose" as the criterion of invalidity and then adds the proviso that legislation will not be invalid, although it violates the prohibition, "if it is necessary to attain some overriding public purpose or to satisfy some pressing social need".[55] This is an echo of Latham CJ in the 1943 Jehovah's Witnesses case discussed above where he asks the question "Does s.116 protect any religious belief or any religious practice, irrespective of the political or social effect of that belief or practice?". And answers it by saying "It has already been shown that beliefs entertained by a religious body as religious beliefs may be inconsistent with the maintenance of civil government. The complete protection of all religious beliefs might result in the disappearance of organized society".[56]

No case has considered in Australia the definition of religious observance as used in section 116. "Religious" may be defined as relating to or manifesting faithful devotion to an acknowledged ultimate reality or deity. "Observance" meanwhile can be defined as a customary practice, rite, or ceremony. It is difficult to see how any law dealing with using embryos, spare or otherwise, in research could be said to be an imposition of religious observance.

Equally it is unlikely that any law banning embryo use would be said to be for the purpose (single or otherwise) of establishing any religion.

This leaves the prohibition of free exercise of religion or its corollary the free exercise of no religion. A number of points can be made.

First, the Federal government has in the recent Report of the Standing Committee on Legal and Constitutional Affairs a clear basis upon which to introduce legislation which *regulates* stem cell and embryo research through a licensing scheme.

Second, the majority Report is prepared to countenance the use of surplus embryos in research. Pausing there, it is difficult to see how such licensing or use of spare embryos could be construed as being an undue restraint on the exercise of either religion or non-religion.

Third, the majority believes there should be a three year moratorium on embryos created by somatic cell nuclear transfer.

Fourth, chapter 7 of the Report is called "The ethics of research and therapy". It deals particularly with the research and application of cloning technologies to human beings. It is asserted that "at the heart of these questions is the degree to which it is ethical to conduct research involving cloning techniques that destroy embryos".[57] Reading this chapter it appears that the significant opposition to the central issue is, at least on the basis of the material quoted in the Report, coming from religious organizations, although citations were made to submissions from individuals. Suggestions were made as to the use of the partial reversal or differentiation of a person's adult stem cells to form regenerative stem cell types. As the Academy of Science said "this is an approach preferred from certain religious viewpoints".[58]

Fifth, the attempt by some to equate their objections not to religious grounds but to a common denominator based, not on religion, but a common humanity, is not conclusive of the matter. Invariably it seems that such submissions come back to a fundamental religious position - that an embryo irrespective of age is to be accorded identical enforceable rights as a person born because an embryo has the essence of being human. The argument is essentially circular.

It is suggested that any law enacted which banned the use of embryos in stem cell research would inevitably, if only in part, be premised on religious objections. If that is so the question then is whether any such law, in giving effect to its purpose of putting in place a ban on embryo research carries within its purpose all of the underlying objections upon which the law is premised. Bear in mind that it is not sufficient that an effect of any such legislation is to inhibit the free exercise of religion or no religion - that must be its purpose. Further, the language of the legislation which is said to offend section 116 of the Constitution must disclose the forbidden purpose.[59] However, it is argued that if a law is enacted, based in part or in whole on certain premises or objections to an existing set of circumstances, then a purpose of the law is to give effect to the objections which form the premise of the law.

One can see scope for judicial use of Gaudron J's "pressing social need or some overriding public purpose" being used in interpreting the envisaged law to ban embryo research. If that is so, then it must be problematical whether a Court would find (without having any specific legislation to consider) that a scientist, in being forced to comply with the ban on using spare embryos or being prevented from creating embryos for the purpose of research, is being forced to accept the religious precepts (which the scientist may find objectionable) and which underlie some of the objections forming the premise of the legislation banning embryo research.

However, the terms of any legislation banning embryo research would need to be closely scrutinized. The cautionary note of Gibbs CJ in *Kerr* "There are limits to the extent to which the law should intrude upon personal liberty and personal privacy in the pursuit of moral and religious aims"[60] should always be borne in mind.[61]

Notes

1 *Human cloning: scientific, ethical and regulatory aspects of human cloning and stem cell research*, Parliament of the Commonwealth of Australia, House of Representatives Standing Committee on Legal and Constitutional Affairs (August 2001).
2 Cabinet decision of the Commonwealth Federal Government reported on 26 February 2002 e.g. *The Age*.
3 The quotes are from an address given by Pope John Paul II to the Pontifical Academy of Sciences and are referred to in Fantoli, A. (1994) *Galileo For Copernicanism and For the Church*, Vatican, pp. 486-7.
4 E.g., Catholic Archdiocese of Melbourne (1999) *Submission to the Commonwealth Inquiry into Scientific, Ethical and Regulatory Aspects of Human Cloning*, Caroline Chisholm Centre for Health Ethics (1999) *Submission to the Commonwealth Inquiry into Scientific, Ethical and Regulatory Aspects of Human Cloning.*
5 Thomson, J. (2000) 'Legal and Ethical Problems of Human Cloning', *Journal of Law and Medicine*, Vol. 8, p. 31.
6 Wallis v Hodson (1724) 2 Atk 117; 26 ER 472.
7 Villar v Gilbey [1907] AC 139 at 144; [1904-07] All ER Rep 779 at 781.
8 Volume IV p. 197.
9 [1972] VR 353.
10 Ibid at 360.
11 Ibid at 376.
12 [1979] QB 276; [1978] 2 All ER 987 at 989.
13 Attorney General (Qld) (Ex rel Kerr) v T (1983) 46 ALR 275.
14 Ibid at 277.
15 F and F (1989) FLC 92-031.
16 See Sieling and Sieling (1979) FLC 90-627 at pp. 78, 264.
17 F and F (1989) FLC 92-031 at pp. 77, 434-5.
18 Ibid at pp. 77, 437.
19 E.g. Seymour, J. (1995) *Fetal Welfare and the Law*; Sullivan, D. (2000) 'State Sanctioned Intervention on behalf of the Fetus', *8 Journal of Law and Medicine 44*.
20 E (Mrs) v Eve [1986] 2 SCR 388 at para 40.
21 E.g. Re X (a minor) [1975] 1 All ER 697 at 699. See also Secretary, Dept Health and Community Services v JWB and SMB [1991-92] 175 CLR 218 at 258 (Marion's case).
22 Secretary, Dept Health and Community Services v JWB and SMB [1991-92] 175 CLR 218.
23 (2000) 50 NSWLR 549.
24 Ibid at 553.
25 Marion's case at 226.
26 (2000) 50 NSWLR 549 at 553.
27 Maw v Western Sydney Area Health Service (2000) 49 NSWLR 231.
28 Ibid at 238.
29 Ibid at 241.
30 Ibid.
31 Ibid.

32 Attorney General (Qld) (Ex rel Kerr) v T (1983) 46 ALR 275 at 277-8.
33 [1979] QB 276; [1978] 2 All ER 987 at 989.
34 Attorney General (Vic) (Ex rel Black) v The Commonwealth & Ors (1980) 146 CLR 559 per Stephen J at 610.
35 Ibid per Barwick CJ at 580-1.
36 Kruger v The Commonwealth [1996-97] 190 CLR 1 per Gaudron J at 131.
37 Official report of the National Australasian Convention Debates (Third Session): Melbourne 2 March 1898 p. 1769.
38 Section 16 Constitution Act 1975 (Vic).
39 Krygger v Williams (1912) 15 CLR 366.
40 Adelaide Company of Jehovah's Witnesses Incorporated v The Commonwealth (1943) 67 CLR 116.
41 Attorney General (Vic) (Ex rel Black) v The Commonwealth & Ors (1980) 146 CLR 559.
42 Kruger v The Commonwealth [1996-97] 190 CLR 1.
43 Krygger v Williams (1912) 15 CLR 366 at 373.
44 (1943) 67 CLR 116 at 123.
45 Ibid at 126.
46 Ibid at 129-132.
47 Attorney General (Vic) (Ex rel Black) v The Commonwealth & Ors (1980) 146 CLR 559 per Barwick CJ at 579.
48 Ibid per Gibbs J at 598; per Stephen J at 609; per Mason J at 615; per Wilson J at 653.
49 Ibid per Mason J at 615-616.
50 Kruger v The Commonwealth [1996-97] 190 CLR 1.
51 Ibid at 40.
52 Ibid at 60.
53 Ibid at 160-1.
54 Ibid at 131.
55 Ibid at 130-4.
56 (1943) 67 CLR 116 at 126.
57 *Human cloning: scientific, ethical and regulatory aspects of human cloning and stem cell research*, Parliament of the Commonwealth of Australia, House of Representatives Standing Committee on Legal and Constitutional Affairs (August 2001) para 7.5.
58 Ibid para 7.52.
59 See Toohey J in Kruger v The Commonwealth [1996-97] 190 CLR 1 at 86.
60 Attorney General (Qld) (Ex rel Kerr) v T (1983) 46 ALR 275 at 277-8.
61 Since this chapterer was written the Australian Federal Parliament has introduced the *Human Embryo Cloning and Research Involving Embryos Bill* 2002 which, if passed, would allow destructive research utilizing human embryos. However, access to embryos will be restricted to those embryos in storage before 5 April 2002 and declared in excess to requirements.

Chapter 20

'Therapeutic Cloning': Tortious Risk Management Issues

Brenda McGivern

Introduction

This chapter explores from a legal perspective the nature of cloning as a therapy. It will examine the issue of whether therapeutic cloning is different in kind from other therapies, and the consequences of those differences from a liability perspective. This chapter has Australian law as its focus, but will have relevance for other common law jurisdictions.

The chapter notes the role of the common law as a regulatory mechanism; one that should not be ignored in spite of the existence of more specific statutory frameworks and guidelines. The focus is on 'therapeutic cloning' which broadly involves the production, by replication, of cells for therapeutic use. This technology encompasses a variety of methods and cell sources and so extends beyond the scope of Assisted Reproductive Technology (ART). However, a cell source thought to offer significant potential benefits in the area is human embryonic stem (HES) cells. Since these cells are derived exclusively from embryos produced ex utero there has been a close association between the use of HES cells, therapeutic cloning and ART.

Significant media attention concerning the controversial use of HES cells in therapeutic cloning has tended to direct attention away from a number of other potentially important issues concerning the therapeutic use of cloned cell products in humans. This chapter therefore notes some of the challenges and issues raised by potential uses of this technology to treat humans.

Part 1: The scientific and regulatory context

Risk management

Legal risk management involves identifying and avoiding sources of legal liability. Essentially, the aim of risk management is to avoid or reduce the incidence of being successfully sued. By understanding the nature and basis of potential liability, one can seek to adopt practices that comply with legal obligations and so avoid being sued. Of course, it is never possible completely to avoid liability. However, risk management has as it goal the adoption of prudent practice that will minimize both harm to patients and liability to practitioners.

It should be clear from the above that the first and essential step is to recognize and outline potential sources of liability. When discussing the regulation of an area of health practice, it is tempting to focus on regulatory mechanisms specific to that area. These usually take the form of statutory regulation. However, legal obligations can arise in a number of ways and this chapter concentrates on potential tortious actions. The law of torts does not, however, operate in a vacuum, so it will be necessary to outline the general scientific and regulatory context in which this discussion arises.

The scientific context

Before we can sensibly talk about liability issues and regulations concerning 'therapeutic cloning', we must first have a clearer understanding of what is meant by that term. This raises a number of definitional difficulties.

Cloning and therapeutic cloning Cloning is a broad term that may be defined as "the production of a cell or organism with the same nuclear genome as another cell or organism".[1] It is therefore broader than the production of an entire organism that is genetically identical to another, and includes the mere replication of DNA and cells. Therapeutic cloning, then, refers to the replication of cells to produce *cells, tissues and organs* for therapeutic use in humans. This is commonly distinguished from so-called 'reproductive cloning' which has as its aim the production of a *child* that is genetically identical to another person. This distinction is itself controversial and can be confusing since both therapeutic and reproductive cloning may involve the creation of a human embryo.[2]

As the definitional discussion illustrates, there is no *necessary* link between therapeutic cloning and ART. There are a variety of potential techniques and cell sources that may be used to produce cloned human tissue for therapeutic use, including:[3]

1. Replication of somatic cells (such as skin).

2. HES cell replication and directed development. Stem cells are relatively undifferentiated cells that retain the ability to divide and produce different specialized cells. Early indications are that this specialization may be influenced and directed, although the capacity to do this is currently in its formative stages. Embryonic stem cells constitute the inner cell mass of a blastocyst (an early stage of embryonic development) and upon their removal the embryo becomes non-viable. Embryonic stem cells appear to offer enormous therapeutic potential as they retain their pluripotent capacity over repeated replications. The other advantage of HES cells over more mature stem cells is that they have been more easily directed to produce particular cell types. HES cells themselves may be derived from two potential sources:
   - 2.1. 'Spare' embryos: these are embryos that are produced for assisted reproduction, but are in excess to the infertile couple's needs. In this instance, the embryo concerned can never be genetically identical to the intended recipient of the cells. Whilst retaining the general advantages of HES cells, these cells (and the tissues they produce) carry the same risk of rejection by the recipient's immune system as is currently experienced with other donated tissues and organs.
   - 2.2. Embryos resulting from somatic cell nucleus transfer: this technique uses the nucleus from one of the intended recipient's own cells to produce an embryo with an identical genetic structure. These cells would be compatible with the recipient, hence avoiding the risk rejection or the need for immunosuppressant drugs.

3. Primordial stem cells (stem cells derived from fetal tissue).

4. Adult stem cells (stem cells derived from adult tissue such as bone marrow).

The link between ART and therapeutic cloning, therefore, exists where the tissue used is derived from HES cells. Since this process requires the extraction of cells at the blastocyst stage of development, the embryo must be created ex utero using techniques employed in ART.

The regulatory context

Whilst this chapter will focus on tortious legal liability, the law of torts must be read in light of the overall regulatory context. The existence of statutory regulation or codes of practice may influence tortious liability. As such, a brief overview of the relevant legislation and guidelines is considered, followed by an outline of the nature of the law of torts and its interaction with statute.

Legislation (public law) There are few regulations that deal with therapeutic cloning specifically. However, certain legislative regulation does exist for cloning *simpliciter*, and this is outlined below. Further, to the extent that therapeutic cloning is reliant on the production of embryos ex utero, aspects of ART legislation are also relevant.

Assisted reproduction is regulated in Australia by a combination of legislative and non-legislative mechanisms. Only three Australian States have legislation that specifically regulates the practice of ART.[4] All three Acts contain prohibitions against cloning.[5] However, the definitions differ under the Acts as to what that term will cover. In Victoria cloning includes the creation of an embryo that is genetically identical to another person or embryo.[6] This would preclude the creation of an embryo as contemplated in 2.1 above. South Australian legislation precludes the creation of two genetically identical embryos from the division of one embryo,[7] and Western Australian legislation prohibits cloning to produce a genetically identical child (i.e. 'reproductive cloning').[8] Thus neither of the latter States expressly preclude the creation of an embryo using somatic cell nuclear transfer,[9] at least where the purpose is not to produce a baby.

The Commonwealth's *Gene Technology Act* also prohibits the cloning of "whole human beings".[10]

Since therapeutic cloning involves not only the creation of embryos, but also their destructive use, the next relevant regulatory issue is the uses to which embryos may be put. The three States with ART legislation place substantial limits on research using embryos.[11] The most significant limitations are those banning research that is destructive or harmful to the embryo.[12] For this reason, research that required the destructive extraction of the inner cell mass of an embryo would be prohibited. The prohibition does not, however, extend to research on embryonic *cells*. It is for this reason that much of the HES cell research currently conducted in Australia relies on the importation of HES cells from jurisdictions without similar prohibitions.

Guidelines In States where no specific legislation exists,[13] Reproductive Medicine Units (RMUs) must comply with the NHMRC *Ethical Guidelines on Assisted Reproductive Technology* (1996) (NHMRC Guidelines).[14] Further, whether or not required by State law, RMUs offering ART must obtain accreditation through a recognized accreditation body.[15] This body is currently the Reproductive Technology Accreditation Council (RTAC). Accreditation with RTAC requires compliance with both the NHMRC Guidelines and RTAC Guidelines.[16] Further, specific research projects or procedures must be approved, approval requiring compliance with both set of Guidelines.[17]

The NHMRC Guidelines provide that research that involves the destruction of an embryo, or that may leave it in a non-implantable state, should only be approved in exceptional circumstances.[18] Further, the list of ethically unacceptable practices to be prohibited[19] include the development of embryos other than for their use in approved ART treatment programs and experimentation with the intent of producing two or more genetically identical individuals.

The RTAC Guidelines include as unacceptable practices reproductive cloning; and replacing the nucleus of a cell of an embryo with a nucleus taken from another person, embryo or fetus.[20]

The non-legislative nature of guidelines means that they may be overridden by State law or a valid Federal law. However, to the extent that no legislation exists, or where it exists but contemplates the operation of the guidelines, these non-legislative mechanisms are an important component of ART regulation. The limitation of non-legislative guidelines, however, is that their enforcement is limited to non-legal means. Typically, Government funding entitlement is effective in inducing compliance. However, since no statutory offence is committed by non-compliance, it is possible for an institution with sufficient independent funding to ignore requirements under guidelines and suffer no public law penalty.

The law of torts (private law) An additional and important regulatory mechanism of health care practice exists in private law actions. These are actions that may be brought by individuals on their own behalf, rather than by and on behalf of the State.[21] The law of torts is a body of law protecting private rights (primarily in one's person or property) from interference by others.[22] Torts law also forms part of the common law. The common law is an unwritten body of law that develops through judicial decision-making.[23] Like guidelines, common law may be overridden by legislation.[24] Thus, where there is a statutory prohibition or requirement, the common law may not permit or require acts contrary to the statute. Similarly, where a statute confers a right or protection, that cannot be removed by the operation of the common law.

Guidelines, on the other hand, do not carry independent legal force and may be subject to common law requirements. Unlike guidelines, torts law operates on broadly recognized rights and duties, rather than from specifically codified obligations.

The largest body of tort case law is that of negligence,[25] but tortious actions include other wrongs, such as trespass, conversion and detinue.[26] Both guidelines and legislation can be relevant in establishing a duty of care, or breach thereof, in a negligence action.

Successful actions in tort sound in financial compensation and can play an important role in regulating health care practices by providing a financial disincentive to engage in practices that risk breach.

Part 2: Potential tortious liability associated with therapeutic cloning

Despite the apparent novelty of some of the techniques utilized in therapeutic cloning, there may be little to differentiate the end product from other biological therapies. When examining the potential application of torts law this area of practice, it is important to analyze the extent to which therapeutic cloning is similar to, or different from, other therapies. In this way, we can extrapolate and predict the application of existing torts principles to novel developments in medical practice.

The areas of potential tortious liability selected for discussion are:

1. Product liability and the duty to inform;
2. Transportation and damage to property; and
3. Reproductive tissue therapy.

In the first topic, we will examine liability issues that derive from the similarity of therapeutic cloning to other therapies. The latter part of the chapter will examine two other areas of potential liability that arise out of some distinctive features that are unusual about the practice of therapeutic cloning.

Product liability and the duty to inform

It has been noted above that the law of torts may be influenced by coexistent legislation. It is often the case that a statute will impose rights and obligations that overlap, but do not extinguish, tortious rights and obligations. Where the common law and statute share certain elements or terms, then each body of law can look to the other to help resolve uncertainties of meaning or interpretation. This has been particularly so in the area of product liability in the so-called "warning" cases.[27]

The law of torts places on a manufacturer of goods a duty of care in respect of users and consumers of those goods.[28] The duty of care is wide ranging and includes a duty to give adequate warning of any risk associated with the use or consumption of the product. This duty has been bolstered and influenced in Australia by the introduction of consumer protection legislation, principally the *Trade Practices Act* (TPA).[29] However, "[t]his statutory change does not supersede the common law and also leaves in place much other legislation governing liability issues arising out of the use of products".[30] Thus, it is necessary to have regard to the operation of the common law and to other relevant[31] legislation.

For present purposes, the question is whether or not the cells, tissues or organs produced using therapeutic cloning can be considered "products". From a liability perspective, it is difficult in principle to distinguish these cloned cell lines from other therapeutic goods, especially pharmacological and biological ones.[32]

The *Therapeutic Goods Act* (TGA)[33] adopts a very broad view of what will constitute therapeutic goods, being goods used for preventing, diagnosing, curing or alleviating a disease, ailment, defect or injury. Further, a limited exemption from the TGA exists in respect of "human tissue for implantation in the human body that is obtained, stored and supplied *without any deliberate alteration to its biological or mechanical properties*".[34] Impliedly, human tissue for therapeutic use that has been deliberately altered will constitute therapeutic goods under the TGA.

The statutory classification of therapeutic goods adds weight to the view that cloned cell lines cannot meaningfully be distinguished from other biological therapies and would be subject to tortious product liability actions. Accepted as such, and in the absence of any specific exemption, the standard of care placed on clinicians producing and supplying the cells would be the same strict standard placed on manufacturers in torts.[35] In the case of warnings, these standards are particularly rigorous.[36]

The concept of imposing on doctors a duty to inform their patients of risks associated with a proposed therapy or procedure is certainly not new.[37] What is significant about classifying the cells produced by therapeutic cloning as "products" is that the duty to inform and warn the ultimate patient will extend back beyond the treating physician to the clinicians involved in the creation (manufacture) of those cells. One ameliorating factor is that, although the manufacturer's duty is owed to the end user, that duty might be discharged by the disclosure of an adequate warning to an intermediate party (for example, from the producer of the cells to the administering doctor). Further, if the 'learned intermediary rule' adopted in North America is followed then the standard of the warning will be sufficient if it enables the prescriber to warn the intended user even if the same warning would be inadequate if directed at the user.[38]

Regarding cloned cells as products means that the chain of responsibility (and hence potential class of plaintiffs) for ensuring the communication of adequate information about the product and its effects will extend to the earliest stages of production and supply. It will therefore be incumbent on those clinicians engaged in their production to consider the liability implications of being a "manufacturer" with "consumers"[39] as opposed to focussing simply on standard doctor/patient relationship and associated liability issues.

Transportation and damage to property

It has been noted that in some jurisdictions where there is a ban on the creation of embryos for research purposes, or on the use of embryos for destructive or harmful research, there are no corresponding prohibitions on the use of embryo-derived cells. As discussed, much of the research in Australia on HES cells is conducted on imported embryonic material. This raises the issue of liability for damage in transit to embryos[40] or embryonic cells. Tortious liability can arise in respect of loss of, damage to, or interference with goods.[41] In the context of transportation, the persons or entities responsible for the transportation of goods are usually bailees of those goods, and are under fairly strict obligations in respect of them.[42]

The question this raises is whether embryos or HES cells are goods for the purpose of recognizing and protecting legal interests in their physical state.

The issue of property interests in human tissue is a vexed and uncertain one.[43] An early body of case law refused to recognize property rights or interests in the human body, or body parts, based on the common law's ancient refusal to recognize property in a corpse.[44] However, early in the 20th Century, the Australian High Court raised an exception to this general rule in circumstances "where a person has by lawful exercise of work or skill so dealt with a human body or part of a human body in his lawful possession that it has acquired some attributes differentiating it from a mere corpse awaiting burial".[45]

Despite the questionable authoritative strength of the exception set out in *Doodeward*,[46] it has since been applied in two recent English cases, *Dobson v North Tyneside Health Authority*[47] and *R v Kelly and Another.*[48] If Griffith CJ's exception were to be applied in Australia, then embryos and embryonic cells created as a result of ART would arguably satisfy the criteria of property. Clearly, the highly technical and contrived creation of embryos ex utero demonstrate considerable "exercise of work or skill". Further, it is certain that these embryos or HES cells acquire attributes that distinguish them from a "mere corpse". However, this latter point raises some uncertainty.

What seems open to question is the significance of the *corpse* to the application of the common law rule or its exception. The issue before Griffith CJ in *Doodeward* was whether preserved fetal remains could constitute an exception to the rule, and this factual background explains the wording of his formulation. However the question remains, does there *need* to be something akin to a corpse for the rule or the exception to apply? The difficulty arises in respect of reproductive materials (gametes and embryos) given their potential for independent life. It might be argued that an embryo is so markedly unlike a corpse that it should not attract the same rules.

It could be argued that most human cells, once removed from the body, are akin to a dead body or its constituent parts. This view explains decisions such as that of the Supreme Court of California in *Moore v Regents of the University of California*[49] that denied a patient's claim to a share of the profits from a cell line developed from his excised tissue.[50] Thus, if the common law rule is understood to apply to *any* human tissue, then Griffith CJ's exception should be read as applying to tissue that acquires through human effort some attributes differentiating it from tissue in its natural state.

Whilst it is appealing to differentiate reproductive tissue from other human tissue, one may question the rationale of doing so in the context of conferring property rights. In fact, there are a number of compelling policy and ethical reasons for refusing to recognize property interests in such tissue. In particular, there is concern about commercializing such tissue and the "danger of promoting the instrumentalization of zygotes, thereby potentially undermining commitment to respect for human life and dignity".[51] However, the policy reasons that support consistent treatment of reproductive materials for the purposes of attracting the common law rule, also support distinguishing them for the purposes of applying the exception. They provide strong arguments for limiting the application of the *Doodeward* exception to tissues that have no potential of developing independent life. That is, even if Australian courts accept and apply the exception, there are sound policy reasons to anticipate that they may not apply it to reproductive material that has the potential to become a whole human being. If this were so, then whilst HES cells in transit might attract the protection conferred by torts law to property, embryos might not.

The ancient origins and uncertainty in this area make it a clear example of the need for modern day legislatures to consider the issues related to and raised by ART and cloning. It is certain that in making its early pronouncements on body parts and corpses, the courts could not have contemplated the uses of human tissue currently being made available by technological advancement. It would therefore be appropriate for Parliament to consider the issue of property interests and the value of therapeutic cell lines, and to develop regulations in that light.

In the absence of specific regulation, the risk of uncertainty remains for parties dealing with and transporting human tissues for therapeutic cloning. One way of managing the uncertainty would be to incorporate specific contractual obligations (and hence liability) in respect of these materials when arranging for their transportation. Given the highly uncertain application of torts law, parties ought to turn their attention to these matters if they are to achieve any level of predicability.

Reproductive tissue therapy

As previously outlined, the use of cloning technologies to create a human baby is widely prohibited and condemned. This is at least partially due to the uncertainty of the long-term consequences of cloning techniques, and the impact that adverse outcomes will have on the resultant child, as well as the potential to replicate those outcomes if the child later reproduces.[52] As such, there has been a strict distinction drawn between therapeutic cloning on the one hand, and reproductive cloning on the other.

In Australia, the Andrews report contains the recommendation that:

> The legislation regulating human cloning and stem cell research contain a ban on cloning for *reproductive purposes*. Any attempt to undertake cloning for *reproductive purposes* should result in criminal penalty and the withdrawal of a license to undertake research in this area for the individual concerned.[53]

However, this distinction would blur if cloned stem cells were used to treat tissue damage to reproductive organs or cells (such as ovarian or testicular tissues). The treatment would still be 'therapeutic' in the strict sense that its immediate purpose or result would be to treat some physical malaise, and would not be the production of a human baby. However, it could (and most likely would) ultimately result in cloned material being used in or for reproduction.

Given the importance of the underlying policy concerns, it is surprising that this issue does not seem to have been addressed in recent reports or debates about cloning technologies. Whilst the phrase "reproductive purposes" used in the Andrews report is arguably wide enough to cover the issue, its terms are not clear and the definitional section of the report makes only the broader distinction between reproductive and therapeutic cloning.[54] Only passing mention is made of this issue in a report commissioned by National Bioethics Advisory Commission in the United States.[55]

Given the current state of stem cell research, the matter is presently hypothetical. However, if such treatment were to become possible and take place, then the potential risk of treatment (and legal liability in respect of it) expands beyond the primary patient (the direct recipient of the cells), and extends to his/her future children.

From a regulatory perspective, the desire to draw a clear distinction between the use of cloning techniques in therapy and reproduction ought to be revisited in light of the potential use of therapeutic cloning in reproductive tissues. If the policy concerns about the safety of cloned tissues used for reproduction remain, then the regulations ought to be reformulated to clarify the extent of any prohibition on reproductive use.

From the health care practitioner's point of view, in the absence of specific regulation, the application of therapeutic cloning to reproductive tissue therapy would broaden the scope of legal risk to include potential actions in tort for wrongful birth, pre-conception injury and, possibly, wrongful life actions.[56]

Conclusion

As illustrated above, when seeking to manage legal risk, both health carers and their legal advisers need to consider both statutory and non-statutory regulatory mechanisms. As science advances it is essential to consider the extent to which the new technology alters or attracts the application of existing legal principle.

Notes

1 This definition was adopted in Australia, House of Representatives Standing Committee on Legal and Constitutional Affairs, *Human Cloning: Scientific, Ethical and Regulatory Aspects of Human Cloning and Stem Cell Research* (Parliamentary Committee Report) (Canberra: 2001) ("Andrews Report") at 19.

2 Ibid at 19; see also Chapters 6 and 7 for discussion of therapeutic and reproductive cloning.

3 For a review of the scientific background, see Andrews Report, op cit n1, Chapters 2-3.

4 South Australia: *Reproductive Technology Act* 1988 (SA) and *Reproductive Technology (Code of Ethical Research Practice Regulations)* 1995; Western Australia: *Human Reproductive Technology Act* 1991(WA); Victoria: *Infertility Treatment Act* 1995 (Vic).

5 Regulations 6, *Reproductive Technology (Code of Ethical Research Practice Regulations)* 1995 (SA); s7 *Human Reproductive Technology Act* 1991(WA); s47 *Infertility Treatment Act* 1995 (Vic).

6 S3 *Infertility Treatment Act* 1995 (Vic).

7 Regulations 1, *Reproductive Technology (Code of Ethical Research Practice Regulations)* 1995 (SA). This process is known as embryo splitting.

8 S3 *Human Reproductive Technology Act* 1991(WA).

9 Both Acts prohibit the replacement of a nucleus of a cell of an embryo or an egg in the process of fertilization. However somatic cell nuclear transfer of the nucleus of an unfertilized egg, and is therefore not caught by the express terms of these provisions.

10 S192B *Gene Technology Act* 2000 (Cth).

11 For a detailed outline of the various provisions see Andrews Report, op cit n1 at 139-145.

12 S14 *Reproductive Technology Act* 1988 (SA); s14 (2) *Human Reproductive Technology Act* 1991(WA); ss24 *Infertility Treatment Act* 1995 (Vic).

13 For a comprehensive review of the non-legislative regulations and their interaction, see Andrews Report, op cit n1 at Chapter 9.

14 *National Statement on Ethical Conduct in Research Involving Humans* (June 1999) (Statement). "Enforcement" is achieved through funding restrictions – lack of compliance with the *Statement* will affect entitlement to funding through the NHMRC.

15 NHMRC Guidelines, s2.1.

16 NHMRC Guidelines, s2.1. The *Code of Practice for Units using Assisted Reproductive Technology* (RTAC Guidelines) requires compliance for accreditation to be approved and retained: "These guidelines are to be observed in clinics or Centres involved in the treatment of patients with ovulation induction, artificial insemination, IVF and related techniques such as gamete intrafallopian transfer (GIFT) and all procedures involving donated gametes or embryos".

17 NHMRC Guidelines, s2.2; RTAC Guidelines, s7.1.

18 NHMRC Guidelines, s6.4.

19 NHMRC Guidelines, s11.

20 RTAC Guidelines, s4.

21 For a more detailed outline of the distinction between public and private law, see Srivastava, DK, et al, *Introduction to Australian Law*, Law Book Company, Sydney, 1996, Chapter 8.

22 See further Davies, M, *Torts*, (3rd ed.), Butterworths, 1999, Sydney; Nygh, P & Butt, P (eds) *Butterworths Concise Australian Legal Dictionary*, Butterworths, 1997, Sydney, at 392.

23 See further Smith, MDH, et al, *Legal Process: Commentary and Materials* (6th ed.), Law Book Company, 1994, Sydney, Chapter 7; Nygh, P & Butt, P (eds.) *Butterworths Concise Australian Legal Dictionary*, Butterworths, 1997, Sydney, at 72-73.

24 Although generally a statute will not be taken to override the common law unless it reveals an express or implied intention to do so. See Nygh, P & Butt, P (eds.) *Butterworths Concise Australian Legal Dictionary*, Butterworths, 1997, Sydney, at 72.

25 Davies, M, *Torts* (3rd ed.), Butterworths, 1999, Sydney, at 3.

26 See discussion below regarding transportation and damage to property.

27 See for example *ACCC v Glendale Chemical Products Pty Ltd* [1998] ATPR 41-632; 40 IPR 619; 180 FCA; *Kato Works Co Ltd v Benz* [1999] WASCA 165; and Malkin I "Product Liability Under the Trade Practices Act and at Common Law: ACCC v Glendale Chemical Products Pty Ltd" (1998) 6 *Torts Law Journal* 204.

28 *Donoghue v Stevenson* [1932] AC 562; *Grant v Australian Knitting Mills* [1936] AC 85; *Black v Lipovac* [1998] 699 FAC; *ACCC v Glendale Chemical Products Pty Ltd* [1998] ATPR 41-632; 40 IPR 619; 180 FCA; *Kato Works Co Ltd v Benz* [1999] WASCA 165.

29 *Trade Practices Act* 1974 (Cth). See *ACCC v Glendale Chemical Products Pty Ltd* [1998] ATPR 41-632; 40 IPR 619; 180 FCA.

30 Luntz, H & Hambly, D, *Torts: Cases and Commentary* (4th ed.), Butterworths, 1995, Sydney at 410.y: Butterworths, 1995) at 410.

31 In this instance, see discussion on *Therapeutic Goods Act* 1989 (Cth), below.

32 There are many instances of biological products being used therapeutically, vaccines being an obvious example.

33 *Therapeutic Goods Act* 1989 (Cth), s3.

34 *Therapeutic Goods Act* 1989 (Cth), s3 and *Therapeutic Goods Regulations* (Amendment) 1992 (Cth), reg 3.

35 See Trindade, F & Cane, P, *The Law of Torts in Australia* (3rd ed.), Oxford University Press, 1999, Melbourne, at 602-624.

36 See Richardson, M "'Towards a test for strict liability in tort': A modified proposal for Australian product liability". (1996) 4 *Torts Law Journal* 24; Malkin I "Product Liability Under the Trade Practices Act and at Common Law: ACCC v Glendale Chemical Products Pty Ltd" (1998) 6 *Torts Law Journal* 204; *ACCC v Glendale Chemical Products Pty Ltd* [1998] ATPR 41-632; 40 IPR 619; 180 FCA; *Kato Works Co Ltd v Benz* [1999] WASCA 165.

37 *Rogers v Whitaker* (1992) 175 CLR 479; *Chappel v Hart* (1998) 195 CLR 232; *Rosenberg v Percival* (2001) 178 ALR 577.

38 Trindade, F & Cane, P, *The Law of Torts in Australia* (3rd ed.) (Melbourne: Oxford University Press, 1999) at 612; see *Hollis v Dow Corning Corporation* (1996) 129 DLR (4th) 609.

39 Recent case law makes it clear that doctors will not be exempt from the requirements of trade and consumer protection legislation. See ACCC v *The Australian Medical Association Western Australia Branch Inc* [2001] FCA 1471.

40 At this stage, the bans on certain uses of embryos would cover embryos created outside of Australia. However, damage to embryos in transit is still a live issue, particularly as the need for transportation can arise for a range of reasons. For example, until recent legislative prevented it, many embryos created in Western Australia were transported interstate to undergo PGD that was not available in Western Australia.

41 Torts actions for loss, damage or interference with goods may arise in negligence (that is, breach of a duty of care) or in intentional torts (trespass, conversion or detinue). For

discussion of these actions see Balkin, RP & Davis, JLR, *Law of Torts*, (2nd ed.), Butterworths, 1996, Sydney, Chapter 4 and Part III.

42 For discussion of bailment and bailees' liability see Fisher, S, *Commercial and Personal Property Law*, (Sydney: Butterworths, 1997), Chapter 5.

43 See generally Stern, K & Walsh, P (eds.), *Property Rights in the Human Body* (London: Centre of Medical Law and Ethics, 1997); Lang, A, "What is the Body? Exploring the Law, Philosophy and Ethics of Commerce in Human Tissue" (1999) 7 (1) *Journal of Law and Medicine* 53; Atherton, R, "Claims on the Deceased: The Corpse as Property" (2000) 7 (4) *Journal of Law and Medicine* 361.

44 See Atherton, R, "Claims on the Deceased: The Corpse as Property" (2000) 7 (4) *Journal of Law and Medicine* 361 at 366-367. See also *R v Lynn* (1788) 2 TR 733; 100 ER 394, a case establishing a criminal offence at common law for disinterment of a corpse without authority despite lack of property interests.

45 *Doodeward v Spence* (1908) 6 CLR 406 ("*Doodeward*") at 414 per Griffith CJ.

46 See commentary in Atherton, R, "Claims on the Deceased: The Corpse as Property" (2000) 7 (4) *Journal of Law and Medicine* 361 at 364-365. The exception was formulated by only one of three members of the bench, each judge finding for the appellant on slightly differing bases. All three, however, did accept the basic proposition that there is no general property interest in a corpse.

47 [1996] 4 All ER 474.

48 [1998] 3 All ER 741.

49 51 Cal 3d 120; 793 P2d 479 (1990).

50 Although the cells were later developed into the cell line through human work, at the time of excision there was nothing to distinguish them from any other body part in which no property was recognized.

51 Canada, Canadian Royal Commission on New Reproductive Technologies, *Proceed with Care - Final Report* (1993) at 638. See also Lynch, MJ, "Property in Human Gametic Material and the New Reproductive Technologies" (1999) 6 *Journal of Law and Medicine* 348; Bennett, B, "The Human Embryo as Property? Cryopreservation and the Challenges for Law" (2000) 7 (4) *Journal of Law and Medicine* 434; Griggs, L, "The Ownership of Excised Body Parts: Does an Individual Have the Right to Sell?" (1994) 1 *Journal of Law and Medicine* 223; Lang, A, "What is the Body? Exploring the Law, Philosophy and Ethics of Commerce in Human Tissue" (1999) 7 (1) *Journal of Law and Medicine* 53.

52 See Andrews Report, op cit n1, Chapter 6. See also United States, *Cloning Human Beings* (Report) (National Bioethics Advisory Commission, 1997); Knowles, L P, United States, *International Perspectives on Human Embryo and Fetal Tissue Research* (Paper commissioned by the National Bioethics Advisory Commission for its report on Issues in Human Stem Cell Research) (National Bioethics Advisory Commission, 2000); United Kingdom, Department of Health, *Stem Cell Research: Medical Progress with Responsibility* (Report from the Chief Medical Officer's Expert Group reviewing the potential of developments in stem cell research and cell nuclear replacement to benefit human health) (London: 2000).

53 Andrews Report, op cit n1, Recommendation 4 at para 12.37, emphasis added.

54 The distinction drawn is between "reproductive cloning to produce a human fetus" and "therapeutic cloning to produce human stem cells, tissues and organs". The definitions adopted were those originally made in Australian Academy of Science, *On Human Cloning: A Position Statement* (Report) (February 1999) at 4.

55 Knowles, L P, United States, *International Perspectives on Human Embryo and Fetal Tissue Research* (Paper commissioned by the National Bioethics Advisory Commission for its report on Issues in Human Stem Cell Research) (National Bioethics Advisory Commission, 2000).

56 For discussion of the nature and basis of these actions see McGivern, B, "Tortious Liability for (Selected) Genetic Harm: Exploring the Arguments" (2002) 10 *Torts Law Journal* 41.

Index